BUILDING a Community

The History of the Orlando-Orange County Expressway Authority

Dr. Jerrell H. Shofner

Copyright © 2001
Orlando-Orange County Expressway Authority

ISBN: 0-9714713-0-4

All rights reserved.

No part of this book may be reproduced or transmitted in any form
or by any means, electronic or mechanical, including photocopying,
recording, or by any information storage and retrieval system,
without permission in writing from the publisher.

Table of Contents

Acknowledgments 4

Chapter 1 5
Origins of the Orlando-Orange County Expressway Authority

Chapter 2 9
Building the Bee Line

Chapter 3 19
Going Downtown: *The East-West Expressway*

Chapter 4 45
Extending the Bee Line: *The Cape Kennedy to Disney World Expressway*

Chapter 5 57
Central Florida's Beltway Vision

Chapter 6 73
The Beltway's Beginning: *The Northeastern Segment*

Chapter 7 87
The Eastern Extension of the East-West Expressway

Chapter 8 95
The Southeastern Beltway: *East-West Expressway to the Bee Line*

Chapter 9 103
The Western Extension of the East-West Expressway: *The Turnpike Connector*

Chapter 10 111
The Road Gets Rough: *The Central Connector*

Chapter 11 121
Building Roads and Mending Fences in the Early 1990s

Chapter 12 135
The Southern Connector, The Southern Connector Extension, and the Seminole County Expressway

Chapter 13 155
E-PASS: *An Electronic Toll and Traffic Management System with Automatic Vehicle Identification*

Chapter 14 171
Privatizing Toll Collections

Chapter 15 181
The Western Beltway

Chapter 16 205
The OOCEA in 2000 and Beyond

List of Abbreviations 219

Bibliography 220

Footnotes 222

Index 238

Acknowledgments

I am indebted to many people for their assistance during the research and writing of this work. Richard Fletcher, OOCEA chairman during the construction of the Bee Line and planning of the East-West Expressway, Phil Reece, who held the same position during planning and design of the 1986 Project, and A. Wayne Rich, chairman from 1992 through 1999, furnished valuable information from their different vantage points. OOCEA staff members Gregory Dailer, Jorge Figueredo, Steve Pustelnyk, and Susan Simon were generous in sharing their knowledge with me. Gregory Dailer and Steve Pustelnyk helped me in locating documents from the vast collections on file in the OOCEA offices.

Executive Director Harold Worrall graciously took time from his busy schedule to help me on numerous occasions. William B. McKelvy and Charles C. Sylvester, who have both worked with the old State Road Department, the FDOT, and the OOCEA since the 1950s, shared with me their vast knowledge of road building in Florida. Gerald Brinton, former executive director of the Seminole County Expressway Authority, helped me to understand the SCEA's role in building the GreeneWay. Donald Erwin, Mike Bierma, Keith Jackson, and Bonnie Eichenberg of PBS&J provided generous assistance.

I am further indebted to Darleen Mazzillo, Vicki Coleman, and Patricia Varela of the OOCEA, and Carla Gregory of PBS&J, for the pleasant manner with which they tolerated my many requests for help.

I also wish to thank Jean Yothers, former curator of the Orange County Historical Museum, for her assistance in obtaining rare photographs.

Jerrell Shofner

Jerrell H. Shofner, professor of history emeritus at the University of Central Florida, is the author of *Nor Is It Over Yet: Florida in the Era of Reconstruction; Daniel Ladd: Merchant Prince of the Florida Frontier; Jefferson County, Florida; Jackson County, Florida; Brevard County, Florida; Apopka and Northwest Orange County; Orlando: The City Beautiful; Florida Portrait* and *A History of Altamonte Springs* as well as many articles in scholarly journals.

He was awarded the Rembert W. Patrick Prize for his *Nor Is It Over Yet* and *Jefferson County*. He is also the five-time recipient of the Florida Historical Society's Arthur W. Thompson Prize and the Frederick W. Weyerhauser Prize for an article in *Forest History*. He lives with his wife, Frances, in Winter Springs, Florida.

Chapter 1

Origins of the Orlando-Orange County Expressway Authority

Orlando and Orange County experienced rapid growth during the decade and a half following World War II. Some of the population increase resulted from the post-war influx of servicemen returning to the salubrious Florida climate which they had discovered during the war.

Others came because of military preparedness relating to the Cold War. Both the Orlando and Pine Castle air bases remained open, bringing payrolls, new businesses, and housing construction. Residential areas spread around the outskirts of the city, and people relied more and more on automobile transportation.

Growth accelerated even further when the United States Missile Test Center opened at Cape Canaveral in 1955. Some of its employees chose to live in Orlando and commute to the coast. Orlando businesses performed services for the new agency. The Martin Company (later Martin-Marietta and then Lockheed-Martin) purchased 7,300 acres in southwest Orange County, spurring more new businesses and residential developments. Martin's 8,000 or so employees commuted from Pine Hills (west), Rio Pinar (east), and Maitland (north). During its first two years, Martin spawned 72 other new businesses. Encouraged by the Orlando Industrial Board, five industrial parks opened in

Looking north on Orange Avenue as it appeared in 1958 during construction of the SunBank Building.

Building A Community

Looking south across Lake Lucerne in 1955 during construction of the Orange Avenue causeway.

the city along with others in surrounding towns. These developments placed commuter traffic on local streets and roads as well as S.R. 50 between Orlando and Cape Canaveral.

By the early 1960s, more than 300,000 people lived in metropolitan Orlando, the economy was still expanding rapidly, and new residents were flooding in. The area's infrastructure, constantly strained, was reaching critical proportions in many areas. Transportation needs were especially serious.

Orlando had been a transportation hub of Central Florida since the opening of the Dixie Highway (U.S. 441) and the Cheney Highway (SR 50) in the 1920s. Progressive business leaders such as Martin Andersen, the influential owner of the Orlando Sentinel, William H. (Billy) Dial, prominent attorney and banker, and others had relished that distinction and were anxious to retain it. With Dial representing Central Florida on the State Road Board, they were able to have the Sunshine Parkway (Florida's Turnpike) routed through southwest Orange County. Interstate-4, connecting Tampa and Daytona Beach, was designed to pass through Orlando and intersect with the Turnpike southwest of town. These roads opened for traffic in 1963 and 1965 respectively. At the same time, area leaders recognized that their Municipal Airport had inadequate runways for jet passenger planes. Pressed for both time and money to construct a new facility, they resorted to extraordinary methods. With Mayor Carl Langford playing a prominent role, they made an arrangement with the U.S. Air Force whereby civilian passenger planes could land at McCoy Air Force Base (formerly Pine Castle) and the city could build a terminal on military land. In 1968, Mayor Langford signed a long-term lease, by which the city paid one dollar to acquire full control of civilian air traffic at McCoy.

With modern airline services after October 1961, and limited access highways to the south, west, and north under construction, Orlando lacked only a similar road to the east. That road became more crucial when President Kennedy announced in 1961 that the United States would put a man on the moon within ten years, leading Brevard County to proclaim itself the "Gateway to the Moon." Wishing to retain its Central Florida leadership position, Orlando changed its own slogan from the "City Beautiful" to "Action Center of Florida," and then took measures to give the new name substance.

A Texan who had kept close contact with his native state, Martin Andersen had long been impressed with what the Dallas Citizen's Council, organized in 1935, had done for the infrastructure of its city. Hoping to provide similar long-range leadership for Orlando, Andersen and others formed the Central Florida Development Commission (CFDC). Its three announced goals were: to promote an area university, to create an East Central Florida Regional Planning Council (ECFRPC), and to complete an adequate road system for the metropolitan area. The ECFRPC was in place by 1962 and Florida Technological University was authorized in 1965. Plans for the road system were more complex.

Lack of funds made traditional road construction methods inadequate. Demands for local roads already far exceeded the resources of local governments. The State Road Department (SRD), grappling with the growing needs of the entire state, was falling behind in its road-building program. The plans for an interstate system of highways, begun in 1956, were already completed, and there were no federal funds for additional limited access highways in the Orlando area.

Faced with these limitations and especially desirous of a limited access road connecting the McCoy Jetport with Cape Canaveral, Andersen and the CFDC explored other alternatives. After a large group of civic leaders visited Jacksonville in 1961 to view that city's expressways, talk in Orlando turned to a toll road system. Drafted by Joel Wells and perfected by Representative John Brombeck who guided it through the legislature, a bill creating an Orlando-Orange County Expressway Authority (OOCEA) became law in 1963.[1]

Although in 1963 the CFDC was immediately interested in building a road between the McCoy Jetport and Cape Canaveral, the law creating the OOCEA authorized an Orlando-Orange County expressway system. Signed into law by Governor Farris Bryant, the statute created an Authority of five members, three of whom were to be Orange County citizens appointed by the governor. The two other members were the incumbent Orange County Commission chairman and the Central Florida member of the State Road Board, serving ex officio. The OOCEA board was to elect one member as chairman. It could elect a secretary and a treasurer who did not have to be OOCEA board members. It was also empowered to employ an executive director, legal counsel, technical experts, engineers, and other necessary employees, and to fix their compensation. It could employ a fiscal agent, but that person or firm was to be chosen from at least three sealed bids. Members were to serve without compensation except for travel expenses incurred in conducting OOCEA business.

The OOCEA was empowered to build expressways and all necessary appurtenances including approaches, roads, and bridges. It could acquire rights-of-way by donation, purchase, or by eminent domain. It also had authority to enter into lease-purchase agreements, to establish and collect tolls, and to borrow money by pledging anticipated revenues. With approval of the County Commission, Orange County tax funds could also be used to back the bonds, but any such funds which were disbursed had to be repaid to the county "when the authority deems it practicable." The OOCEA was also authorized to accept grants from, and to enter into contracts with, any federal agency, the state, the county of Orange, or the city of Orlando.[2] This was a broad grant of authority to carry out a large responsibility. When need for roads were great and no tax dollars were available, the OOCEA would become the builder of last resort.

Although it received scant notice at the time, a proviso that the OOCEA could not build a road within the boundaries of any municipality without the municipality's approval was to have broad repercussions many years later.[3] Another limitation was that the OOCEA lacked power to build outside Orange County, and the need for metropolitan roads often extended beyond the county boundary.

In the meantime, the OOCEA was first concerned with constructing the road which became known as the Bee Line.

Chapter 2

Building the Bee Line

The Orlando-Orange County Expressway Authority had an inauspicious beginning. Meeting in borrowed quarters in early July 1963, charter members Raymond E. Barnes, O. P. Hewitt, Lloyd Gahr, Orange County Chairman Donald Evans, ex officio, and Max Brewer, State Road Board member, ex officio, organized for business. Barnes was elected chairman with O. P. Hewitt as treasurer. Vera Gordon was employed as secretary. Subsequent meetings were held at various places, including the county commission offices, the Cherry Plaza Hotel, and the offices of several engineering firms. From 1965, when Richard Fletcher became chairman, through mid-1967, the OOCEA was housed in his insurance agency. A budget of $10,000 was adopted for the first year. It was funded by loans of $5,000 each from Orange County and the city of Orlando. The funds were deposited with the Florida National Bank and the secretary kept accounts.[4] With a tiny budget funded by borrowed money from a desk loaned by the SRD and located in donated office space, the OOCEA began planning its first road project.[5]

The modest circumstances were misleading. The agency was charged with responsibility for building an expressway system for the burgeoning metropolitan Orlando area. In 1965, the ECFRPC was predicting the seven county area around Orlando would have a population of 1.8 million by 1980. The city alone was expected to have a population exceeding a million by 1990. With high expectations and the enthusiastic support of Orange County, the city of Orlando, the SRD, the CFDC, the ECFRPC, the Orlando Area Chamber of Commerce and even the Seminole County Commission, the OOCEA

Raymond E. Barnes, first chairman of the OOCEA. (Photo courtesy of the Orlando Sentinel.)

O. P. "Pete" Hewitt, charter member of the OOCEA.

Building A Community

Donald S. Evans, Orange County Commission chairman and charter member of the OOCEA, serving ex officio.

Max Brewer, State Road Board member and charter member of the OOCEA, serving ex officio.

immediately went to work. At the first meeting, it was made clear that the Bee Line, to connect I-4 with Cape Canaveral, was only the first step. The OOCEA was also looking beyond it to other limited access roads. There was even discussion of a "beltline" around the metropolitan area.[6]

As a basis for initial planning, the OOCEA asked the Florida Turnpike Authority (FTA) for a copy of its recently completed study of road needs between Orlando and Cape Canaveral. It then selected the firm of Coverdale and Colpitts to make a traffic study of such a route and accepted a $50,000 loan from the SRD to pay for it. From a list of 16 engineering firms, the OOCEA selected Reynolds, Smith and Hills, and Howard, Needles, Tammen and Bergendorff, both with Orlando offices, as their consulting engineers.[7]

Wishing to build a limited access road all the way from I-4 southwest of Orlando to Cape Canaveral, the OOCEA searched diligently for ways to finance such a large project. It had authority to sell revenue bonds based on traffic studies showing a reasonable expectation that toll collections would amortize the incurred obligation. It could also ask Orange County to pledge the first $450,000 of its secondary gasoline tax funds to back the bonds. But Orange County was authorized to make such a pledge only if the traffic projections were adequate.

The OOCEA looked for ways to add to its revenue base by including Brevard County, which certainly had an interest in the proposed road. But, there were two major problems. First, while many Brevard County residents saw the value of such a road, others regarded it as Orlando's effort to draw people away from their retail businesses and to make the McCoy Jetport the airport for the space program. There was also some belief that Martin Andersen had used his influence with national leaders to delay completion of I-95 in favor of I-4. That belief was hardly assuaged by signs which appeared in Volusia County advising motorists to "avoid U.S. l" and take I-4 to the Turnpike for a better route to South Florida.[8]

A second and more concrete problem was in the law that created the OOCEA. Even if Brevard County wished to participate, its gasoline tax revenues could not be pledged for use by the OOCEA whose scope was limited to the confines of Orange County. The OOCEA's attorney, Robert Ogden, advised that a change in the law would be necessary if Brevard County were to be included in financing the road.[9]

The complications of including Brevard County in the project were soon replaced by more immediate woes. The engineering firms had been busy planning the best routes. By early 1964, three plans were being considered. Plan A, by far the most favored, called for a limited access road connecting I-4 and Cape Canaveral. Its estimated cost was $36,000,000. When Coverdale and Colpitts presented their traffic study, it showed that anticipated use of the road, at rates ranging from 35 cents for passenger cars to 80 cents for the largest trucks, would not come even close to justifying such a large bond issue.[10]

OOCEA Chairman Sidney Singleton, who had succeeded Raymond Barnes upon the latter's death, convened a meeting at which the board discussed the next step. Assuming that their traffic count would support a bond issue of about $7,000,000, they decided to see what could be built for that amount. Max Brewer, the State Road Board member, suggested that Plan C, for a road entirely within Orange County, be considered. That was all that was necessary "for the present," he said.[11] His colleagues agreed and turned to the more limited and more attainable plan.

Even Plan C, for a road between I-4 and S.R. 520 (popularly referred to as the Bithlo Cutoff) required considerable trimming to fit into anticipated available funds. The final product was to be part controlled access expressway and part toll road. Using Sand Lake Road as the I-4 connection, that road would be extended eastward from U.S. 441 (Orange Blossom Trail) past Orange Avenue and the Atlantic Coastline Railroad tracks to McCoy Road. Next, the controlled access expressway would continue from Daetwyler (the entrance to McCoy AFB) to S.R. 15 (Narcoosee Road). Then, the limited access Bee Line would extend 17.4 miles from a toll plaza at S.R. 15 to S.R. 520.

While the engineers worked on the specific design and route of the pared down road, the OOCEA turned to the lengthy and complicated process of

Sidney Singleton, OOCEA chairman, 1963-1965.

Summary of the bond document for the Bee Line project prepared by J. Fenimore Cooper, Jr. (OOCEA attorney) and the OOCEA Board.

financing it. With the favorable traffic study in hand, the OOCEA turned to the County Commission for a pledge of the first $450,000 of its secondary gasoline revenues to back the bond issue. The traffic study showed that 3,000 cars a day at 35 cents toll would produce enough revenue to service the debt, but the county's backing would increase the marketability of the bonds.[12]

A well-attended public meeting in May 1964 convinced the county commissioners that their constituents approved the pledge of funds. With the exception of two political candidates and the owners of businesses along S.R. 50 east of town, there was vigorous favorable reaction. Especially well-received was Max Brewer's assurance–in response to a question–that the purpose of the OOCEA was not just one expressway. The Bee Line was just one segment.[13]

When the OOCEA began drafting its bond resolution in June 1964, a member asked how long it would take to obtain financing and complete construction. He was told that it would take six months to prepare and sell the bonds and then 18 months to complete the road. The questioner lamented that that meant two more years for a road that "is needed now."[14] The sense of urgency continued, but the estimate was too optimistic. To prepare the bond resolution, obtain the county's pledge of gasoline funds, get approval of the SRD, the State Board of Administration, and the Bond Review Board, validate the bonds, and then advertise and sell them, required more than two years.[15] While the members toiled over the wording of a trust agreement with the bond purchaser, Attorney J. Fenimore Cooper, Jr. was asked why it was so important. He replied "because we will have to live with it for 40 years."[16]

The bonds were sold in November 1965 to the First Boston Corporation at a rate of 4.06 percent, and the trust agreement worked quite well.[17]

Since most of the route was through undeveloped territory, right-of-way acquisition was not too difficult. The road from U.S. 441 to McCoy Road (about 2.3 miles) was built on Orange County-owned land. From Daetwyler to S.R. 520 there were about 20 properties. Nelson Boice, a local realtor, had been working as a volunteer on acquisition, and by September 1964 had obtained oral commitments on about 80 percent of the route. Some of the land was donated and one rancher even added $20,000 toward construction of a bridge on his property. There were to be four such bridges along the way to permit the movement of cattle on the grazing land and, perhaps, later to provide places for roads to cross when the area was developed. There was another advantage of this method because the Martin Company was planning to use the Bee Line to transport large shipments, and an unobstructed road would save both time and inconvenience.

To transform Boice's oral commitments into contracts, the OOCEA employed Ray Johnson as right-of-way agent along with two appraisers. They arrived at fair prices and, in most cases, formalized the agreements which Boice had made. Right-of-way acquisition, including appraisals, cost only a little over $300,000 for the entire project. Only four properties had to be acquired by condemnation. The property which caused the most difficulty, and which was ultimately settled without condemnation, was a large tract just east of S.R. 15 owned by J. J. Brunetti of Ft. Lauderdale. The OOCEA was still dealing with him when construction started on the Bee Line, forcing the contractor to begin work far out in the flatwoods and build in both directions.[18]

Another necessary step was completed in November 1964, when the OOCEA signed a lease-purchase agreement whereby the SRD would take over operations and maintenance of the Bee Line when construction was completed and remit toll revenues to the OOCEA for debt management.[19]

Building A Community

Willard Peebles succeeded Max Brewer, who died in a plane cash in 1964, and served as ex officio member of the OOCEA board from 1965 to 1969.

There were several personnel changes by the time construction began in early 1966. Richard L. Fletcher began his long tenure as OOCEA chairman in early 1965, succeeding Sidney Singleton. Max Brewer, the State Road Board member, was succeeded in 1964 by Willard Peebles. Donald Evans was replaced by John Talton as ex officio member from the Orange County Commission. The last charter member of the board was O. P. Hewitt who was replaced in late 1965 by E. C. Goldman of Winter Park. J. Fenimore Cooper, Jr. had already succeeded Robert Ogden as OOCEA attorney, and Victor F. Wasleski became the first executive director in August 1966.

The final design of the Bee Line and its connectors called for several separate projects to be built concomitantly. The largest was the Bee Line itself, a 17.4 mile limited access toll road from S.R. 15 eastward to S.R. 520. There were to be four bridges for cattle crossings and another across the Econlockhatchee River. The road bed was to be raised four to five feet along the entire route. The 4.5 miles between Daetwyler and S.R. 15 was another project. A third was the 2.3 miles between U.S. 441 and McCoy Road with a bridge over the Atlantic Coast Line railroad tracks. The toll plaza constituted a fourth project. There was additionally a widening project of S.R. 15 from

Map of original Bee Line project, including the widening of S.R. 15 between Conway and Goldenrod roads.

14

its intersection with Conway Road to its intersection with S.R. 15A. Although not part of the OOCEA's building program, the State Road Department was preparing to extend Lake Barton Road (S.R. 436) from Curry Ford Road to the new highway.

The main Bee Line project was awarded to Hubbard Construction, a longtime road builder in Orlando, whose low bid was $3,247,875. With the other smaller projects, two of which Hubbard also built, the total cost of construction was a little more than $5,000,000.[20]

Nearly five years had passed since the CFDC made its trip to study the Jacksonville expressway system and more than two years had elapsed in securing financing when road construction finally began in 1966. But, an elated Richard Fletcher still promised completion of the road in 18 months, less time than had been spent in preparation. There was extensive publicity and excitement when ground-breaking ceremonies were announced, but the location assured that the audience would be small. George Daniels, newly elected vice chairman of the OOCEA, led a caravan of engineers, contractors, and reporters along a tortuous route, including eight miles of dirt road to Wewahootee, then several miles along a trail across George Terry's Magnolia Ranch, and one and a half miles along an abandoned Florida East Coast Railroad bed to a stretch of pine and palmetto flatwoods where three bulldozers awaited them. There, in the Florida flatwoods surrounded by miles of grazing land, the first shovel of earth was turned to start construction of a project which Norman Bryan of Reynolds, Smith and Hills declared had "been a long time coming."[21]

Ground-breaking ceremony for the Bee Line. In center (5th from left) is George Daniels, vice chairman of the OOCEA, speaking with Norman Bryan of Reynolds, Smith and Hills. The four men on the left are (from left) H. E. Lewis, Don Gage, Eddie Marvin and Bob Trux. The two on the right (from left) are Art Beard and W. C. Peterson. Gage, Trux, Marvin, and Beard were Hubbard Construction employees. Lewis and Peterson were with Reynolds, Smith and Hills.

With equipment such as this, Hubbard Construction removed 500,000 yards of muck and replaced it with dirt suitable for the Bee Line road bed.

Preparation of the road bed for limerock and asphalt.

Construction proceeded without major interruptions. Hubbard lost about 100 days to wet weather, but a long period of what Richard Fletcher called "contractor's weather" allowed the firm to regain most of the lost time. In the spring of 1967, Fletcher was still predicting an opening by July 1. An anxious public waited while workers hurried to help Fletcher keep his word. Although they were not quite successful, the Bee Line was dedicated at the newly constructed toll plaza on July 14, 1967. The remaining portion of the road was opened nine days later.[22]

In December 1966, the OOCEA had voted to name the road the "Martin Andersen Bee Line Expressway." The legislature enacted legislation at its 1967 session. On July 14, Richard Fletcher acted as master of ceremonies as Willard Peebles cut the ribbon opening the 17.4 mile toll road for which passenger vehicle drivers would shortly be paying 35 cents for the privilege of using. The audience was considerably larger than the one which had attended the groundbreaking in the Florida flatwoods 18 months earlier. It heard Peebles extol Martin Andersen's role in obtaining the Bee Line as well as his promotion of Central Florida during the preceding 30 years. Seemingly embarrassed for having bestowed upon him an honor usually reserved for the deceased, Andersen commented that he was afraid someone was going to say "Doesn't he look natural?"[23]

It was appropriate that Andersen be recognized for his leadership role, but many others had worked hard to make the event possible. Richard Fletcher, whose own efforts had been considerable, told his OOCEA colleagues in late June that they had built 23 miles of four-lane road at a cost of $280,000 per mile, a feat which he predicted would not soon be repeated.[24] Drivers could exit I-4 at Sand Lake Road and drive a straight road for 23 miles at

George Daniels (left), Eldon C. Goldman, F. B. Surguine, and Orlando Mayor Robert Carr reading the resolution naming the Bee Line for Martin Andersen.

The first brochure published by the OOCEA to advertise the Bee Line.

a cost of 35 cents, saving about 13 miles as compared to the old route to Cape Canaveral.[25] The total outlay for completion of the road was ultimately placed at $6,800,000.

Fletcher was justifiably proud of the OOCEA's accomplishment, but Max Brewer's earlier explanation that the Bee Line was just a beginning was already being demonstrated. In 1965, even before construction began, Walt Disney had announced plans for his immense theme park. That announcement immediately transformed public perception of the Bee Line. It soon became

Richard L. Fletcher speaking at the dedication ceremony for the Bee Line on July 14, 1967. Martin Andersen is seated to his right wearing a white jacket. On his right are Eldon C. Goldman, Jay Brown of the SRB, and James T. Cooper. The others are unidentified. (Photo courtesy of the Orlando Sentinel.)

Martin Andersen is receiving the first ticket for travel on the Bee Line from M. G. Blozak. On the left are Charles Rex of the Florida Turnpike Authority and Richard Fletcher. Willard Peebles is barely visible on the right. (Photograph courtesy of the Central Florida Regional History Center.)

Building A Community

Looking west along the Bee Line toward the Orlando Airport in the early 1970s. The International Corporate Park exit, the state prison, and the Stanton Energy Plant would be built later in the upper center.

referred to as the Cape Kennedy to Disney World Expressway and, even before the original project was completed, plans were being discussed for extending both ends of the Bee Line to provide a limited access route from Cape Canaveral to I-4 near Disney World.

Then, in 1966, Governor Haydon Burns asked the OOCEA to look into the possibility of an expressway across downtown Orlando from east to west to relieve the worsening congestion on S.R. 50 (Colonial Drive). While the SRD and the FTA began working on the Bee Line extensions, the OOCEA turned its attention to the East-West Expressway.

This photograph is also looking west farther along the Bee Line in the early 1970s. The S.R. 15 (Narcoosee Road) interchange is in the center, and the airport is clearly visible in the upper left.

This view is east from I-4 along Sand Lake Road about 1967 when it was the western segment of the original Bee Line. The interchange just east of I-4 is for the Martin plant. (<u>Photograph courtesy of the Orlando Sentinel.</u>)

Chapter 3

Going Downtown:
The East-West Expressway

The need for an East-West Expressway across Orlando was becoming increasingly obvious, not only to the planners who were anticipating the enormous future growth of the metropolitan area, but to the drivers who were finding S.R. 50 (Colonial Drive) more and more overcrowded by the mid-1960s. The road was carrying about 5,000 more cars per day than it had been built to handle, and the number was increasing rapidly. Noting that I-4 was providing a limited access route for north-south traffic, critics were pointing out that Orlando had an east-west dimension, too. All that traffic depended on S.R. 50, the only east-west road across town.

Unlike the Bee Line, which had been built through largely undeveloped land, an expressway across Orlando would require disruption of miles of urban development, with accompanying expense for right-of-way. There were also the logistical problems of relocating displaced families as well as dealing with heavy construction in a populated area. But something had to be done to keep Orlando from choking on its own traffic. There was some discussion of double-decking S.R. 50, but the many disadvantages of such an action and the estimated $150,000,000 cost were compelling reasons to look elsewhere. A more appealing form of relief was introduced in June 1965, when the ECFRPC published a report recommending construction of a limited access highway through the city. Walt Disney's announcement of his plans for a huge theme park southwest of town and the continuing expansion of National Aeronautics and Space Agency (NASA) gave the ECFRPC report added emphasis. And there was more. The Naval Training Center was expected to open in 1968 with Florida Technological University to follow in 1969. If S.R. 50 was overcrowded in 1965, what would it be like in a few years with Disney World, the Naval Training Center, and the new university putting more cars on the roads?

Building A Community

Governor Haydon Burns 1965-1967. A strong supporter of improved roads, Burns urged the OOCEA to build the East-West Expressway. (Photograph courtesy of the Florida State Archives.)

Governor Haydon Burns, whose experience as mayor of Jacksonville, made him aware of the value of good roads, wrote OOCEA Chairman Fletcher in early 1966 strongly urging action. He wrote, in part, that he was aware that the "demands of planning, design, and construction" of the Bee Line were nearing completion, and that "I believe the Authority should immediately initiate plans for an East-West Expressway to serve Orlando. I would appreciate early action . . . and periodic reports of the progress you are making."[26]

Fletcher and his OOCEA colleagues accepted Governor Burns' strong recommendation and acted. Willard Peebles, the State Road Board member of the Authority, assisted in obtaining a loan of $25,000 for the requisite feasibility studies. Since Wilbur Smith and Associates of New Haven, Connecticut, had already spent more than a year on a traffic survey of metropolitan Orlando for the SRD, they were asked to do the

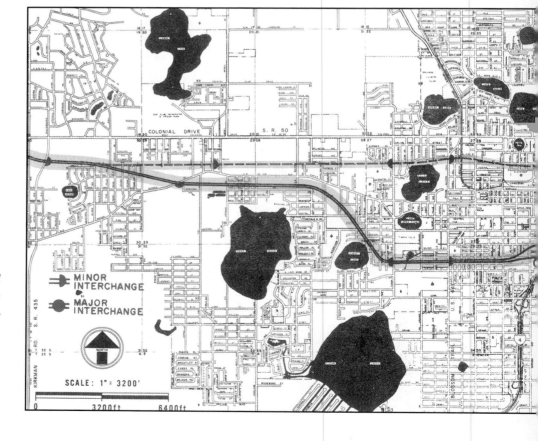

Consulting engineer HNTB and Reynolds, Smith and Hills recommended two possible routes for the East-West Expressway. Alternate one is shown here by the dashed line which crosses and then recrosses S.R. 50. Alternate two is the southern route which was selected.

traffic study. Norman Wuestefeld, representing Wilbur Smith, said the firm could report in four months whether potential traffic would support a toll expressway across Orlando. A preliminary engineering study was a joint venture by Howard, Needles, Tammen, and Bergendorff (HNTB), and Reynolds, Smith and Hills.[27]

The engineers recommended two possible routes. The northern route, alternate-one, would start near Powers Drive and parallel S.R. 50 eastward past Magnolia, cross over, and continue on the north side to S.R. 436 (Lake Barton Road), cross over again, and rejoin S.R. 50 east of S.R. 15A (Goldenrod Road). It would be 12.9 miles long and would cost $82.5 million. Of that sum, about $39.1 million represented right-of-way costs. The southern route, alternate-two, started at the same point and turned south near the business district to run along the south Anderson Street corridor, crossing Lake Underhill, S.R. 436 and S.R. 15A, then turning northward to rejoin S.R. 50 just west of Union Park.

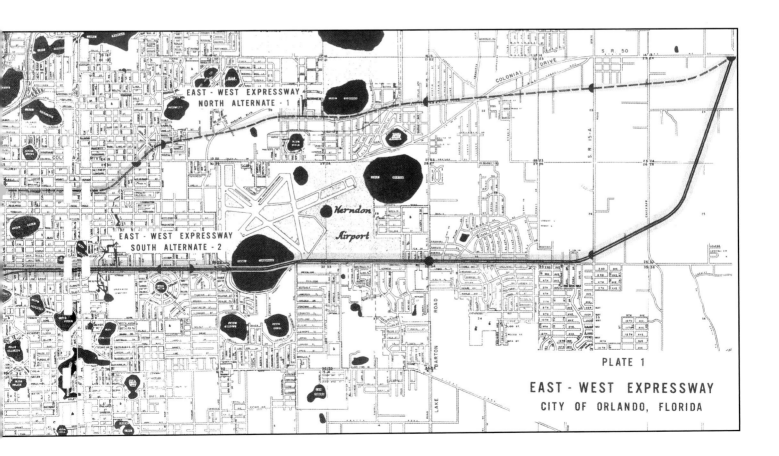

PLATE 1

EAST - WEST EXPRESSWAY
CITY OF ORLANDO, FLORIDA

The distance was 13.3 miles and would cost $58.7 million. Estimated cost of right-of way for this route was $24.7 million.[28]

The OOCEA realized that staying as close as possible to S.R. 50 would maximize traffic on the new road, but the greater right-of-way outlay and the more costly construction for the northern route made it financially prohibitive according to the Wilbur Smith traffic analysis.[29]

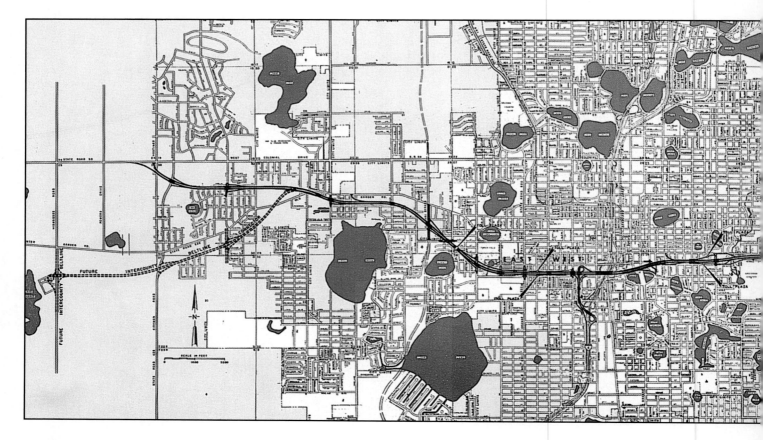

The East-West Expressway as finally designed. Note in far left of the photograph where Hiawassee Road was then considered to be the "future intercounty beltline."

In a wide-ranging discussion about how and whether to proceed, it was suggested that the legislature "might come up with" a new statewide bond issue. C. H. Peterson of HNTB said Orlando might get the road built with federal interstate funds "if you can wait until 1983." Fletcher said that the philosophy of the OOCEA was that, when tax dollars were unavailable from any source for needed roads, the OOCEA was the builder of last resort. In keeping with that philosophy, it was decided to proceed with the less costly southern route along the south Anderson Street corridor, while seeking the

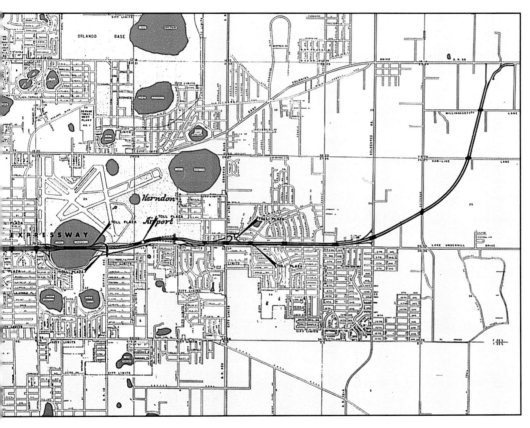

financial assistance of Orange County. Such assistance was a strong possibility since the Bee Line traffic was predicted to be adequate to manage that debt. Fletcher stated that the engineering reports under discussion were exploratory only, and specific details, such as whether or not it was possible to avoid a causeway over Lake Underhill, would be more closely examined.[30]

When the SRD report on metropolitan Orlando's traffic projections was released in the spring of 1968, showing a need for $500 million in new construction, it was agreed that the East-West Expressway should be built as a toll road.[31] In July, with the help of Willard Peebles, the SRD loaned the OOCEA an additional $250,000 to complete the necessary engineering studies. The loan was to be repaid when bonds were sold for expressway construction.[32]

Harry Bertossa of HNTB was named general consultant. The firms of Reynolds, Smith and Hills, Watson and Company, and Voorhies, Trindle-Nelson (VTN) were retained as engineers for specific sections of the road. Bertossa said the work could be completed in about six months so that the OOCEA could use it to proceed with the bond validation process. At the same meeting, John S. Rushing was named Executive Director.

A jubilant Willard Peebles said that after the planning was finished, construction could be completed within three years. Since Disney World was scheduled to open in late 1971, such a schedule was highly desirable. But Peebles' optimistic estimate was not to be realized.[33]

Not everyone was as happy as Peebles. While many residents along the selected route agreed that new roads were necessary, some of them were outraged about the possibility of losing their property. Hearing that many streets would be closed, some denounced the new road as "the great wall" which would divide the City Beautiful. Fletcher and other OOCEA members were criticized for their decision. In a retrospective letter, Martin Andersen subsequently thanked Fletcher for his courageous action in the face of "bitter criticism from people who did not know and would not understand."[34] While some people simply did not want to move, nearly everyone in the affected area was anxious to know if and when his property would be taken. Others wanted to know where they could go. The long delay between initial planning and the acquisition of right-of-way, though unavoidable, caused considerable anxiety. The OOCEA later took measures to ease the plight of those displaced, but some members subsequently recognized that it might have done a better job of informing the public.

While the OOCEA had extensive powers, it was not entirely autonomous. It was dependent upon state and local governments in several ways, and there

Colonel John Rushing (second from left) served as OOCEA executive director from 1968 to 1972.

were many political and constitutional changes afoot in the late 1960s which would affect the progress of the East-West Expressway. One of them, an extensive reorganization of state government required by the new 1968 Constitution, abolished the old SRD with its five-member board and replaced it with the Department of Transportation to be headed by a single secretary. With his position abolished, Willard Peebles would no longer be able to provide his valuable conduit to the state's road building authority.

There were also political changes. In a surprise election, Haydon Burns had been ousted from office in 1966 to be succeeded by Claude Kirk, the first Republican governor of Florida since the 1870s. This was causing considerable partisan change at all levels of government even as the pugnacious new governor battled with a Democratic legislature long-accustomed to one party government. One result was delay in filling positions created by the reorganization mandated by the new constitution.

At the same time, the East-West Expressway was a far larger and more complicated undertaking than the Bee Line had been. It was ultimately decided to close 58 streets and build bridges over 37 others. There were about 1,250 properties, including nearly 1,100 homes, about 80 businesses, six churches, and some vacant land in the path of the proposed expressway. This meant not only the complexity of obtaining right-of-way from people being obliged to leave their homes, but also lucrative employment for real estate agents, property appraisers, attorneys, and building movers. Given these amorphous conditions, Peeble's prognosis was not likely to be realized. In fact, the abolition of his position was one of the first changes under the new regime to be noticed.[35] Wishing to assure some continuity, Governor Kirk did reappoint both Fletcher and E. C. Goldman whose terms were expiring in 1969. Fletcher was reelected chairman and Goldman continued as vice chairman.[36]

By June 1969, plans were calling for a limited access road generally along the southern route announced earlier. It would be four lanes on each end and six lanes in the downtown area. Some details of the exact route were still being

Governor Claude R. Kirk, 1967-1971. The first Republican elected governor since the 1870s, Kirk succeeded Haydon Burns in a surprise upset. (Photograph courtesy of the Florida State Archives.)

worked out, and land owners were becoming increasingly anxious to know how they would be affected. Fletcher said, "If everything goes the way we want it, we could sell a $64.5 million bond issue (inflation was already taking a toll) in March 1970, let contracts in July, get construction going, and be open in the fall of 1972." He admitted that his prognosis was "iffy."[37]

In order to avoid delays, the OOCEA in August 1969 sought an additional $750,000 loan, this time from the new Florida Department of Transportation (FDOT), to proceed with pre-design and preliminary first phase work, such as field surveys, boring contracts, soil testing and utility work orders. The loan was not soon forthcoming. The governor and the legislature were embroiled in a lengthy struggle over appointment of FDOT personnel. While Disney was employing 3,000 carpenters, as well as other workers, and anticipating a 1971 opening, the East-West Expressway project waited.[38]

Not until February 1970 did the FDOT Acting Secretary Jack Cashin even respond to the loan request, and then he rejected it because it was "an awful lot."[39] Another delay resulted from confusion over appointment of a fiscal agent. Goldman Sachs & Co. was finally named in December and had just begun to work on the necessary bond issue. Since significant inflation was adding an estimated $300,000 per month to the cost of the project, Fletcher and his colleagues were understandably disappointed at the delays.

Newly appointed Secretary Edward Mueller finally approved the loan in May 1970 along with a $70.5 million bond issue. The cabinet approved the bonds for validation in October. Preliminary engineering work was nearing completion and property appraisers were already in the field working on the 1,250 parcels to be acquired for right-of-way. Of course, no contracts could be offered until the bonds were sold.[40]

The new constitution provided for pledging "the full faith and credit" of the state in support of such bond issues if the county first pledged its secondary gasoline tax funds. Such a guarantee was certainly desirable since it would make the bonds more attractive to potential buyers, but unfortunately it was being challenged in court in a case involving a bridge contract in Escambia

County. The OOCEA's bonds could not be offered for sale until that case was decided. The Supreme Court finally upheld the provision in early 1971, the year that had once been projected as the completion date for the East-West Expressway.[41]

The bonds were sold in May 1971 to a syndicate formed by Smith, Barney and Company at 5.925 percent, a favorable rate for the time.[42] Commercial Bank of Winter Park had already been named trustee for the $70,500,000 in a vote on November 11, 1970.[43] When asked why it had taken five years to get ready for construction, Fletcher replied that there had been a constitutional change and major reorganization of state government, three governors, three secretaries of transportation, and a major court case. But he might have added that, without the belated $750,000 loan which Secretary Mueller had approved in May 1970, the delay could have been even longer. That loan had enabled completion of preliminary engineering plans and appraisal of 50 percent of the right-of-way before the bonds were sold. That shaved approximately six months from the ultimate completion date of the Expressway.[44] Expectations were to begin construction in early 1972 and to open the road in October 1973.[45]

About the time the crucial Supreme Court decision was handed down, three new members joined the OOCEA. James B. Greene and John H. Talton were appointed by recently inaugurated Democratic Governor Reuben Askew, and County Commission Chairman Ralph Poe became a member ex officio.[46] The two holdovers were Fletcher and Goldman. Fletcher, who had led the OOCEA through completion of the Bee Line and the planning and financing of the East-West Expressway, even housing the OOCEA in his business office from 1965 to 1967, did not wish to continue as chairman and had said so in writing. But he was surprised at the ensuing events.

While Fletcher was out of town, the new members held a meeting at which he was replaced as chairman by Ralph Poe, and John Talton was elected treasurer. Goldman, the only dissenter to the hasty action, remained vice chairman. The next item of business was to fire J. Fenimore Cooper of Cooper and Gilman, who had served as general counsel since 1964. He was immediately replaced by Dan Hunter, the incumbent mayor of Winter Park and a member of the firm of Hunter, Pattilo, Powell and Carroll. A storm of criticism followed. Some argued that Poe was too new to the OOCEA to take over at such a critical point in the sale of the East-West Expressway bonds, and almost everyone disapproved of the hurried methods by which he had obtained the office. There was even greater animosity over the firing of Cooper and hiring of Hunter. The Orlando Sentinel was quick to point out that Poe and Hunter were close friends, and that impending legal work relating to the OOCEA's acquisition of 1,250 parcels of land would be quite lucrative for the attorney handling them.[47]

James B. Greene, OOCEA chairman from 1971 to 1984 with only a brief interruption. During his tenure the East-West Expressway and the OIA-Bee Line Improvement projects were completed, and plans were begun for the 1986 Project. (Photograph courtesy of the Orlando Sentinel.)

In early March, a majority of the Orange County Commission passed a resolution instructing Ralph Poe to call a meeting of the OOCEA and reorganize it. Poe eventually agreed to comply, although he, of course, could not control the actions of the other members. At a March 4 meeting, Dan Hunter, in a statement highly critical of the Orlando Sentinel, resigned as general counsel. Poe then resigned as chairman, but retained his position as ex officio member representing Orange County. James Greene was then elected chairman, a position in which he served with distinction until 1984.[48] Except for the unseemly manner by which he was replaced, Fletcher had no objection to the outcome. Both he and Poe served out their terms as contributing members of the OOCEA.

A longtime Orlando resident, the 33-year-old Greene had a successful insurance business, had been active in public affairs, and was ready "to get going on this project as soon as possible."

His immediate goals were to complete the bond sale, finalize the engineering plans and address the problem of spanning Lake Underhill to the satisfaction of all parties. At a meeting in late March, noting that much work lay ahead, Greene called on all members to become involved by accepting specific assignments. Talton was placed in charge of administration, Poe was responsible for engineering, Fletcher would handle finance, and Goldman's specialty was construction.[49] John Rushing's duties as executive director were expanding, and Harvey Gaines was busily building a staff to handle right-of-way once the bonds were sold. Other staff members would be employed as the need arose. At the suggestion of Angus Barlow of Ernst and Ernst, auditor for the OOCEA, a mechanized bookkeeping system was installed to handle the impending volume of transactions.

Harry Bertossa of HNTB would continue as general engineering consultant during construction and afterward until final details of the huge project were completed. Dale Patten and John Kelly, also of HNTB, were project engineers.

There were continuing requests for changes in the design of the road. The more serious of these had to be worked out so as to stay ahead of construction requirements as well as to keep within the budget. A serious problem of this nature arose while the bond sale was still pending. The City of Orlando wanted assurances that planned interchanges at Herndon Airport and the Tangerine Bowl-Tinker Field area would be upgraded. The OOCEA said that would be done "when funds are available."

The city had agreed to turn over more than $2 million worth of land it owned to be used for right-of-way. The OOCEA agreed to compensate the city for it, if there were funds remaining after the bonded debt was paid.[50] In February 1971, the city demanded additional changes, each of which would have cost money the OOCEA did not have. Harry Bertossa explained that the requested improvements could be made in the future when traffic increased and funds were available. The Orlando Central Business District, Inc., asked for a bridge rather than fill between Lake Lucerne and Silvia Lane so that Garland Avenue would not be blocked. There were no funds for the change at the time, but the FDOT subsequently provided the $370,000 necessary to build the bridge. When the city dropped its objections to the expressway, Jim Greene praised the city councilmen for their action saying, "It is not the Expressway Authority building the road. It is a partnership between the City of Orlando, Orange County and the Department of Transportation."[51]

There were also serious objections to the expressway design at Lake Underhill. Plans called for a 60-foot-wide strip to be filled about 600 feet on the west bank and about 500 feet on the east with a bridge in the center. Estimating that the fill would take about 10 million gallons of primary water storage, the Orange County Water Advisory Board called for a bridge across the entire lake. Opponents also complained that the silt and future runoff would destroy the lake's water quality and pose a hazard to recreational activities on the lake. Fletcher explained that a bridge spanning the entire lake would add $3,000,000 to the costs of the expressway, a sum which the OOCEA did not have. As for the silt and subsequent runoff, barriers would be set up to contain the silt during construction and runoff would later be channeled to pass through filters before entering the lake.[52] Still dissatisfied, Suzanne Buie and Bruce Barnhill, acting for the Sierra Club, filed suit for an injunction prohibiting the city from selling land near Lake Underhill necessary for the expressway, which Barnhill said, "We definitely need . . . There's no question about that." But he and Buie were suing because the road would deprive them of their right to a clean environment. The suit was eventually rejected.[53] The expressway was built as originally planned and Lake Underhill is still a popular recreational area.

A change in design which the engineers made in 1970 to reduce costs caused unfavorable publicity later. Because the area was then sparsely populated, it was decided that bridges could be eliminated and grade level crossings permitted at three streets in the Chickasaw Trail area, saving about $4.4 million. The bridges could be added later. Subsequent accusations of collusion with private landowners were raised and then disproved.[54]

As soon as the bonds were sold, right-of-way activities accelerated. Harvey Gaines continued his work on appraisals. Each property was appraised by two people and acquisition agents were instructed to use the higher figures. Kenneth Ketelson was placed in charge of acquisitions. He and seven other agents began making purchase agreements with property owners whose appraisals had been completed.[55] They first concentrated in the areas between

Cottage Hill and Parramore, between Parramore and I-4, and on the east from Mercado Drive to S.R. 50. The first construction was to begin in these three areas. Those agreeing to settle for the appraised amount could be paid on closing in about 60 days. In the cases of property owners unwilling to accept the appraisal price, condemnation procedures would be implemented with the court setting the price at a later date. The acquisition agents also offered sellers relocation assistance.

There had been no plans for helping people to find new housing, but when it became clear that displacement would work severe hardship on at least 300 families, Greene and his colleagues decided that something had to be done. With housing prices rising rapidly in the early 1970s, the number of hardship cases was likely to rise. In what became the first non-federal project in Florida to offer financial assistance to people displaced by road construction, the OOCEA established a relocation program for which the FDOT eventually authorized an expenditure of up to $2,000,000. Since it was illegal to pay relocation allowances, persons needing assistance would be paid "moving allowances." Harvey Gaines was responsible for the program, and it was managed by Phillip D. Steinmetz. Two field offices were opened at 607 W. Carter Street and 519 E. Anderson Street where families could be helped with their housing needs. O. R. Colan and Associates of West Virginia was employed to assist in the relocation program. Cooperating were the Orlando Area Chamber of Commerce, FHA's local office, Orlando Neighborhood Improvement Program, Orlando-Winter Park Board of Realtors, the Home Builders Association, Orlando Housing Authority, and Orlando Federal Savings and Loan Association.[56]

Given the large number of families – about 410 home owners and 700 renters – being displaced, and the sharply rising prices at the time, the relocation program was quite successful. By January 1972 – the time when construction was beginning – the workload had diminished so that the Anderson Street office was closed and three acquisition agents were terminated. In March, only 250 families were still living in the right-of-way,

This brochure was distributed to the more than 1,100 families and individuals whose dwellings were in the East-West Expressway right-of-way in an effort to explain what was happening and why it was necessary.

Many houses had to be demolished. This is the remainder of a demolition project near Lake Lucerne in late 1971.

and many of them were merely waiting to occupy houses that had been bought or rented, or were being built. Gaines reported in April that acquisition was completed except for "a few odds and ends." Everyone had been relocated by November 1972. The relocation program had cost about $1.6 million. Despite considerable apprehension and some legitimate complaints, Harry Bertossa said that in his 20 years of experience, this had been the "smoothest urban right-of-way acquisition" he had witnessed.[57]

This house is being moved from the right-of-way and relocated.

Tom Bressani was the property manager who received the parcels as they were acquired, cleared them, and turned them over to the engineers. Many of the houses were to be demolished, but especially in view of the housing shortage in Orlando, some were suitable for relocation by firms possessing the equipment and skills to do so. Parcels destined for clearing were advertised for bids in groups as they became available. A tract of 30 units in Azalea Park was the first advertised for bids in July 1971. Others followed fairly quickly until 11 firms were selected to clear the right-of-way. The largest tract comprised 600 buildings for which Dore Wrecking Company of Orlando was paid $292,861.[58] Nearly all the parcels were cleared and turned over to the engineers before they were needed for construction. According to William J. Bates, who had succeeded Bressani as property manager, there were only 23 structures left in the entire corridor by mid-August 1972.[59] About $23.5 million had been expended for right-of-way acquisition.

This was the first multi-story structure removed from the right-of-way. The four-unit apartment complex was cut into two 100-ton sections and moved from 45 Carter Street to 2300 W. Central Blvd. by George Saunders of Building Movers, Inc.

The remaining properties were in the area between Bumby Avenue and Lake Underhill Drive, which was under contract for construction by Diversacon, Inc., of Winter Park. Building Movers, Inc., had paid the OOCEA $3,599 for about 59 houses in the area between Bumby

Avenue and Lake Underhill Drive, intending to relocate most of them. The contract completion date was May 1, 1972. The firm had already missed that deadline when it hired subcontractors from Ohio and Georgia to help complete the work, and the OOCEA agreed to extend the contract until July 1.[60] Other extensions followed until Diversacon Vice President Gerald Hardage notified the OOCEA in November that his firm was being delayed and that he intended to ask for compensation. The problem was finally resolved on December 21, nearly seven months after the original contract deadline, but "still pending [was] the matter of final resolution of the overrun of contract time."[61] With that exception, involving only one of Diversacon's three contracts, all right-of-way was cleared and turned over to the contractors in time for construction to begin.

This open space downtown has been cleared of houses and is being prepared for construction.

Although numerous firms would be involved in building the expressway and opening it to traffic, it had been decided to divide the heavy road construction into seven parts, enabling seven major contractors to proceed simultaneously. Realizing that demands for concrete beams, concrete and asphalt for paving, and perhaps other materials would exceed the capacity of local suppliers at any one time, there was an effort to stagger the work to avoid that problem. But the success of the strategy was dependent upon many factors, including coordination of right-of-way acquisition and clearing, obtaining the mammoth supply – an estimated eight million cubic yards – of required dirt, and management capability of the several construction firms, labor supply, and weather. An eighth large contract for installation of toll facilities was let much later.

Between January 5 and August 14, 1972, seven contracts were completed amounting to about $41 million. The first went to Ballenger and Company of

Governor Ruben Askew speaking at the "ground-building" ceremony for the East-West Expressway. Seated behind him from the left are Secretary of Transportation Edward Mueller, James Greene, and Eldon C. Goldman.

South Carolina in conjunction with Structures, Inc., to build the portion between Mercado Drive and east S.R. 50 for $5,861,000. Diversacon, Inc., of Winter Park was awarded a contract on January 7 to build in the I-4 area, including the crucial interchange there. The amount was $6,464,000. In March, Cone Brothers of Tampa, the low bidder for the portion between Parramore and John Young Parkway, was passed over, and a contract was let instead to Hubbard Construction of Orlando to do the work for $5,688,000. Cone Brothers sued, but the OOCEA's decision was upheld. On May 18, Wiley N. Jackson of Roanoke, Virginia, was selected to build from I-4 eastward to Summerlin Avenue, including the long bridge over Lake Lucerne, for $5,206,000. On June 2, a second contract was awarded to Diversacon to build from Lake Underhill eastward to Mercado Drive, including the controversial bridge over the lake and two others. That contract was for $7,700,000. The sixth contract was awarded to Cone Brothers to build from Cottage Hill Road westward to S.R. 50 for $4,174,000. The final contract also went to Diversacon to build between Summerlin and Lake Underhill for $6,162,000. This gave Diversacon a huge portion of the entire road, including the complex interchange at I-4 as well as the controversial bridge over Lake Underhill, a large undertaking for a single firm. But if all the contractors could keep their schedules, the suppliers of building materials would probably not be overburdened and the October 1973 opening of the expressway could be met.[62]

In December 1971, just as the OOCEA began advertising for bids on expressway construction, it voted to name the road the "Spessard Lindsay Holland East-West Expressway," in honor of the popular four-term United States Senator from Lakeland who had just retired from office. The legislature confirmed the action at its next session.[63] Emphasizing Jim Greene's

The Metropolitan Life Insurance Building at 525 South Magnolia was saved from destruction to become the central office of the OOCEA.

commitment to the environment, a "ground-building" – rather than the traditional "ground-breaking" – ceremony, was held on February 5, 1972, at 5729 Lake Underhill Road in the foundation of a demolished home. Governor Reuben Askew, Mayor Carl Langford, and other dignitaries placed shovels of earth around the roots of a 30-foot magnolia tree to mark the beginning of construction.[64]

A series of administrative and personnel changes occurred in early 1972. Having been housed at 121 Court Avenue for several years, the Orlando-Orange County Expressway Authority moved into its permanent quarters at 525 South Magnolia on May 12, 1972. Just prior to the relocation, Executive Director John Rushing announced his resignation. He said that administrative work on the East-West Expressway was about finished, and he did not wish to be involved in the Outer Beltline project for which the OOCEA was beginning to plan. He was succeeded by Harvey Gaines, who had managed the complex right-of-way project. A few months later, Elmo Hoffman of Hoffman, Hendry, and Parker was named general counsel in place of longtime OOCEA attorney J. Fenimore Cooper, who had asked to be replaced and be named special counsel. It was not long, however, before Cooper resumed the position.[65]

Work on the expressway was getting underway when the "ground-building" ceremony was held. Although they probably caused no more inconveniences and conflicts than any urban public works project of such size, the moving of eight million cubic yards of fill dirt, the frequent closing of streets, and the permanent elimination of an exit ramp on I-4, caused many complaints, some of them quite heated. Union Park Fire Chief James Dunham received a flood of complaints about the smoke resulting from mountains of burning debris near that community, but that episode did not compare with the fury resulting from Diversacon's hauling of fill dirt along Forsyth Road from a Heather Street borrow pit to its Lake Underhill area project.

The OOCEA's new office building awaiting occupancy while the East-West Expressway is being completed overhead.

Dirt from this borrow pit south of town was being hauled on Forsyth Road to the East-West Expressway construction site.

In August 1972, with a permit from Orange County, Diversacon's subcontractors began hauling dirt in 70,000 pound trucks along unpaved Forsyth Road where residents were soon complaining about noise, dust, mud, house-rattling vibrations, and safety hazards for their children. Although the company tried to live up to its contract by frequent grading and watering of the road to smooth out the ruts and keep the dust down, tempers had reached the boiling point by November. When several drivers quit because they had been struck by rocks or other missiles, Sheriff Mel Coleman increased patrols in the area, but the dispute raged on. Responding to complaints, the Orange County Commission, despite the permit it had issued, passed an ordinance to reroute the trucks. A circuit judge threw it out. Faced with a $1,500 a day penalty for failing to meet contracted schedules, Diversacon went to 24-hour a day hauling in December. Irate residents threatened violence. With the Sheriff obliged by law to keep the road open, and the residents condemning the county officials for abandoning them, the dispute raged until Diversacon completed that hauling project.[66] The firm still had more hauling to do, but had agreed to use another route, along Forsyth Road north to S.R. 50, to S.R. 436 and south to Lake Underhill. The respite was short-lived for Forsyth Road residents, but their new adversary was another firm which had its own difficulties with Orange County.

By December 1972, projectiles were being thrown, obstacles were being placed in the road, and drivers were quitting.

The Ballenger Corporation and Structures, Inc., were building the easternmost leg of the expressway between S.R. 50 and Mercado Drive. They had contracted with the county to clean out several drainage canals and use the dirt for fill. But two problems emerged. The county was late in securing right-of-way to the property – condemnation proceedings were still in court – and then, the company discovered that the county had promised the owners that the dirt would be spread on their land. To avoid falling farther behind on its schedule, Ballenger's subcontractor had secured its own borrow pit to make up for that loss from the canals. The trucks, which kept Forsyth Road residents upset after Diversacon left, were hauling for Ballenger.[67]

The problems of maintaining a tight schedule – which the OOCEA had in mind when it staggered its major contracts – was compounded by the huge number of subcontractors upon which the prime contractors were in turn dependent. Those working for Ballenger and Structures, Inc., were exemplary. Grading and excavating was the responsibility of Leecon and Craggs & Phelan who, in turn, contracted with S. E. Montgomery Trucking. Other subcontractors were Wilson Construction Co. – bridge work; Houdaille-Duval-Wright – pre-stressed beams for the bridges; Grasshopper, Inc. – grass and sod; Sloan Construction Co. – asphalt paving; J. E. Hill – guard rails; O'Brien Construction Co. – box culverts; Bethlehem Steel – reinforcing steel; and Orange Concrete – ready mix concrete. And there were six other projects of comparable size underway.[68]

Construction in progress on the west end of Lake Underhill bridge.

Misener Marine driving pilings for the bridge over Lake Underhill in August 1973. Some of them were 220 feet long.

Having been fully aware of the environmental concerns regarding the 1,701 foot crossing of Lake Underhill, Diversacon exercised great care in surrounding the fill with a polyethylene "diaper" reaching to the lake bottom at depths ranging from five to 25 feet. Although the public never became involved in that portion of the work, placing pilings for the bridge posed another problem for Diversacon and its subcontractor, Misener Marine Construction, Inc. With an estimated 250 feet of muck below the lake and the first few feet being very soft, placing the pilings was tricky. But Misener worked it out. Ranging in depth from 80 feet to 220 feet, and without touching bed rock anywhere, the pilings were driven to support 200 tons.[69]

Other problems which arose from time to time caused delays, but none were so controversial as the Forsyth Road hauling matter. Except for the complications relating to the I-4 interchange, none were so difficult as the Lake Underhill bridge, and Diversacon also had to deal with that one.

Piers being driven for the East-West Expressway bridge over I-4 late 1972.

When the terms of Richard Fletcher and E. C. Goldman expired in January 1973, Governor Reuben Askew appointed Richard Swann and William Poorbaugh to succeed them. They joined Jim Greene, John Talton, and Paul Pickett, ex officio from Orange County. At a February 12 meeting of the OOCEA, Talton asked for the floor, said he thought it would be best if the chairmanship be rotated, and nominated Swann. With only Greene voting no, Swann was elected chairman.[70]

Continuing with business, Swann referred to the "preliminary" selection of an engineer for the proposed Outer Beltline and declared that no one had actually been chosen. Because he and Poorbaugh were new members, he wanted to review the engineers for the job. Actually, HNTB had been selected for the work in October 1971. This was the firm which employed Harry Bertossa, general consultant for the East-West Expressway, as well as Dale Patten and John Kelly, the project engineers, all of whom had been involved in the controversy with Building Movers, Inc., in late December.[71]

The swift chairmanship change and the ensuing move to reopen the search for a consulting engineer for a position which had already been filled set off a serious controversy. FDOT Secretary Walter Revell fired off a demand that Swann resign and return the chairmanship to Greene. He said all state funds – a $150,000 loan for the preliminary study of an Outer Beltline – would be suspended until "we can meet with the people involved."[72] Revell, of course, did not have authority to force the personnel change, and the majority of the OOCEA refused to act. While stories circulated about how the action had come about, Governor Askew announced that he had personally asked that Swann resign and that Greene be reinstated. He told the Sentinel that he disapproved of the ousting of an experienced chairman in favor of a newcomer to the board.[73]

Although he lacked authority to remove the members without cause, Askew called on them to reinstate Greene or resign. William Poorbaugh resigned, but the remaining members met on February 26, ostensibly to interview engineering firms for the consulting job on the proposed Outer Beltline. None of the invited firms appeared, probably because FDOT Secretary Revell had

Governor Ruben Askew, 1971-1979. The governor's micro-management of the OOCEA restored Jim Greene to the chairmanship, but it had serious side effects. Not only did William Poorbaugh resign rather than accede to the governor's demand, there were also calls for abolition of the OOCEA. But the most serious effect was the collapse of the interlocal agreement which Greene had only recently completed with Orange, Osceola, and Seminole counties.

announced that he would not approve any change to the selection already made. He later asked the OOCEA to proceed as the FDOT's agent in making a study of the Outer Beltline and select HNTB to do the work.[74]

The Sentinel Star reaction to the OOCEA controversy and the governor's interference.

A letter from Richard Swann was before the board saying that he was resigning because the governor had called for the reinstatement of Jim Greene, and that the OOCEA needed the support of the governor. John Talton then took the floor and said that in view of the governor's request, he wished to nominate Greene for chairman. The motion carried. Shortly afterward Askew appointed J. Thomas Gurney, Sr., a 72-year-old Orlando attorney who was well-known for his long public service, to succeed Poorbaugh.[75]

The strife which gripped the OOCEA during most of February did little to interrupt the building of the East-West Expressway, but it had disastrous repercussions for the proposed Outer Beltline. As Governor Askew had said in mid February, "Jim Greene has just concluded the very difficult assignment of negotiating agreements with three counties – Orange, Seminole and Osceola – and with the city of Orlando."[76] As mentioned earlier, the OOCEA had authority to act only within Orange County, yet a beltline around the metropolitan area needed to cross into both Seminole and Osceola counties. In a lengthy effort, Greene had forged an agreement with all agencies involved to plan the road around the metropolitan area. At the time of the disruption, the agreement lacked only the FDOT signature to become official. Afterward, all three counties withdrew their support with Seminole specifically citing the "squabble" as the reason.[77] State Representative William L. Gibson even declared his

Looking northwest across Lake Lucerne and causeway construction in April 1973.

intention to initiate legislation abolishing the OOCEA upon its completion of the East-West Expressway.[78]

Concentrating on the chairmanship controversy, the news media gave little attention to the end of Richard Fletcher's eight years of dedicated public service on the OOCEA, but others took notice. Willard Peebles wrote that, "I will always associate the name of Richard Fletcher with the Holland Expressway." Complimenting Fletcher for having withstood angry criticism while performing a great service, Martin Andersen wrote, "Yours is a great achievement. A public trust honorably and completely fulfilled."[79] Asked about these and other laudatory letters, Fletcher modestly replied, "We saw it as a duty, believed in getting the job done, and then getting out so that others could continue on."[80]

Although there were no budgeted funds for it, Jim Greene was calling for landscaping along the expressway as early as 1971 and asking the city to remove as many shrubs and trees as possible and store them for use later. First calling the program the East-West Parkscape, he called for volunteers to plan a landscape which would blend with the communities through which the road was passing and give drivers a pleasant view. Prominent local architect Nils Schweitzer led a three-day meeting in October 1972, at which 80 volunteer architects and landscape planners devised a plan and presented it on a 120 foot runner. To raise an estimated $200,000 needed to implement it, Greene spoke to every public service organization in the county which would listen. This shoe-string initiative led to an excellent landscaping

Looking west from the Parramore neighborhood on October 25, 1973. Orange Blossom Trail is in the center. Holland West Toll Plaza is in upper right.

Looking west on the East-West Expressway on October 25, 1973. Old Winter Garden Road is to the right.

program. Harry Bertossa reported in 1976 that it was the only work still in progress on the East-West Expressway.[81]

Inflation and overruns drove the cost of the road beyond the amount available for it. There were several changes in the plan to allow for additional exits and other minor changes, removal of unanticipated muck beds cost millions, and relocation of utilities was more expensive than originally expected. While frugal investments of OOCEA funds earned about $4.5 million to help defray extra costs, the FDOT ultimately advanced slightly less than $15 million to cover the shortfall. The road's final cost was about $89 million.[82]

By September 1973, construction delays had produced the problem the OOCEA had sought to avoid. Contractors were all needing asphalt, concrete, and rock at the same time, and suppliers were falling behind. The project farthest from completion was Diversacon's in the downtown area. As opening date approached, it became clear that the entire road would not be ready. On October 26, 1973, the western half of the road from S.R. 50 to Mills Avenue was opened. Construction crews worked through the night of October 25 to finish that seven mile section. To mark the opening, OOCEA members and city, county and state officials inspected the new highway by motorcade. The full route was open by December 11, 1973, although work continued on portions of it until April 1974. Drivers were charged 20 cents at the two main toll plazas and 10 cents at the ramp exits.[83] According to the lease-purchase agreement, the FDOT took over operation and maintenance responsibilities for the road.

The western terminus of East-West Expressway and S.R. 50. Kirkman Road is at the center of the picture.

Motorcade marking the opening of the western portion of the East-West Expressway on October 26, 1973.

Harvey Gaines (right), shown here speaking with Orlando Police Chief Robert Chewning, supervised the acquisition of right-of-way for the expressway. In 1972, he succeeded John Rushing as OOCEA executive director.

Initial results were disappointing. Daily collections were far short of the projections upon which the bond issue had been based. The oil embargo of 1973 and the ensuing recession were partially responsible for the lack of use. Other possible causes were the delays in the connection of John Young Parkway to the expressway and the extension of Mills Avenue to connect it with U.S. 17-92 leading to Winter Park and Maitland.

Despite considerable criticism, though, traffic increased annually. An advertising campaign had an immediate effect. Some of the problems with the collection equipment and personnel furnished by the FDOT were resolved. While the East-West Expressway was not supporting itself, excess proceeds from the Bee Line and loans from 80 percent of the Orange County gasoline taxes enabled the OOCEA to manage its debt. The road was self-supporting by 1978. A daily average of 73,000 cars used it that year, and interest payments for both it and the Bee Line were made on January 1, 1979, leaving a surplus of more than $600,000.[84]

By 1976, the OOCEA staff had been reduced to Executive Director Harvey Gaines and Executive Secretary Myrtle Rizza. With an annual budget of $60,000, its duties were to record and deposit daily tolls and invest the money in U.S. Treasury securities. The board members had not had a meeting between January and July of that year. The Outer Beltline was in limbo because of fuel shortages ensuing from the oil embargo and the accompanying downturn in the economy.

Myrtle Rizza, shown here at her retirement party in 1978, worked with Richard Fletcher on the Bee Line project and with James Greene on the East-West Expressway. In 1976 she and Gaines were the agency's only two employees.

Building A Community

This brochure was published in 1974.

For the second time in less than four years, the Sentinel Star called for abolition of the OOCEA.

The Sentinel Star, citing these points, called on the legislature to abolish the OOCEA. Such an action, the paper added, should be accompanied by a vote of thanks to Greene and the other OOCEA members who had served without pay to produce two excellent roads which "will stand as a monument to their work."[85]

The dismissal was premature. The OOCEA still had an important role to play. Soon after the Sentinel Star called for its abolition, it embarked upon plans which would keep it busy until the year 2000.

Chapter 4

Extending the Bee Line:
The Cape Kennedy to Disney World Expressway

Before the first shovel of earth was turned on the original Bee Line project, Walt Disney's celebrated announcement of his plans for Disney World changed public discussion of the road. It immediately became "The Cape Kennedy to Disney World Expressway," and the portion then under construction in Orange County became "the bob-tailed Bee Line," or "the mid-section." In November 1965, two months before that construction began, the Orlando Sentinel wrote that "it is a great stroke of happy fate that the [Bee Line] project is well underway," because "this new road when completed will provide a safe and fast route for the millions of tourists who will come to see the two greatest attractions in the state and nation."[86]

Predicting that 10 million tourists would be visiting Walt Disney World® by 1975, Robert Doyle of the ECFRPC recommended completion of the limited access highway so that some of them might more easily visit the NASA attractions. He called on Orange and Brevard counties to conduct a new traffic study taking into account the anticipated Disney crowds. Both counties were receptive. After years of opposing the Bee Line, Brevard County leaders finally realized that the road ran both ways.[87] The Brevard County Commission and the OOCEA agreed on a joint study to determine the feasibility of a limited access road from I-4 to U.S. 1 in Brevard County. Willard Peebles, the State Road Board member of the OOCEA, promised to get it started soon. Doyle urged that the study include the possibility of dividing the road in Brevard County with a fork leading north to the Orsino Causeway (S.R. 405) to serve the NASA tourist center, and another curving south to connect with the Bennett Causeway and the southern gates of Cape Kennedy, Port Canaveral, Cocoa Beach, and Patrick AFB.[88]

Charles W. Rex, Jr., chairman of the Florida Turnpike Authority. (Photo courtesy of the Orlando Sentinel.)

General discussions about the proposed new road were still underway in early 1967 when FTA Chairman Charles W. Rex, Jr. announced his plans for a major expansion of the Sunshine State Parkway (Florida Turnpike) which would include lateral toll roads connecting with the main road. Working in close cooperation with the SRD, Rex said the FTA would, among other things, purchase the existing Bee Line and extend it on both ends to connect I-4 with the Atlantic Coast. The eastern extension would include a fork connecting with the Orsino Causeway and another leading to the Bennett Causeway. The FTA would also purchase the Bennett Causeway and four-lane it from its Bee Line connection to Port Canaveral. Rex apparently intended to make the connection with I-4 by way of an intersection at the Turnpike near its Orlando-South exit, and then north on that road to its intersection with I-4. The plan was contingent upon legislative approval. Rex was preparing to ask the 1967 legislature to grant that authority and to approve a $93 million bond issue to finance that and other new construction.[89]

Having already discussed the plan with Rex, OOCEA Chairman Fletcher gave his hearty approval, noting that it would relieve the OOCEA of its outstanding debt and strengthen its ability to finance the impending East-West Expressway through Orlando.[90]

The requisite legislation was enacted with little opposition and Governor Claude Kirk signed it into law in early July 1967.[91] Rex predicted that the bonds would be sold, construction begun, and traffic would be rolling on the new road in 1970. He did not take into account fiscal inflation, petty politics, constitutional changes, governmental reorganization, and the dusky seaside sparrow, all of which would contribute to the delay of completion of the Bee Line extension until 1974.

By the time the FTA was prepared to market its bonds, inflation had driven interest rates above the five percent allowed by law. Both Rex and Jay Brown of the SRD said the joint effort by their two agencies to complete the road in time for Disney World's 1971 opening depended upon legislation raising the interest rate limit. Despite considerable support in the legislature for such legislation, Representative Clifford McNulty of Melbourne blocked the measure from being voted out of committee. McNulty, representing south Brevard County, whose interests differed from central Brevard (Cocoa) and north Brevard (Titusville), had earlier opposed the 1967 legislation authorizing the Bee Line extension to those sections of the county.[92] Proponents of the road were obliged to look for alternative financing.

Charles W. Rex (left), James T. Cooper, Orange County Commission chairman; Richard Fletcher, chairman of the OOCEA; Lee Wenner, Brevard County commissioner; and Jay Brown of the SRB, pointing to the map of the proposed Bee Line extensions upon which they had just agreed. (Photo courtesy of the Orlando Sentinel.)

Putting their heads together in a cooperative spirit which surprised some observers, the SRD, the FTA, the OOCEA, the Brevard County Commission, and the Orange County Commission worked out a plan. The FTA authorized the use of $6 million of its surplus funds to build the western extension from McCoy Jetport to the Turnpike at its Orlando-South exit. Using a combination of SRD primary funds, income from tolls, and pledges of secondary road funds, the other agencies made available $33 million to build the two-pronged eastern extension. A $10 million bond issue would be shared by Brevard and Orange counties to build the Bee Line eastward with a northern extension to the Orsino Causeway. Brevard County would issue another $23 million worth of bonds, a portion of which would be used to four-lane the Bennett Causeway to the mainland and extend it to connect with the Bee Line.[93]

The bonds were sold through the Florida Development Commission which had them ready for market in December 1968. They were sold to a syndicate assembled by Smith, Barney and Co. at six percent, the maximum allowed by the state and the highest rate in state bonding history to that time.[94]

Intersection of the Bee Line and S.R. 436 in July 1969. Construction of S.R. 436 is nearing completion. In the foreground is the old terminal of Orlando Jetport. (Photo courtesy of the Orlando Sentinel.)

Looking southwest across the intersection of S.R. 436 and the Bee Line toward the Orlando Jetport terminal in the early 1970s. It was replaced by a multilevel interchange as part of the 1980 Project to serve the new Orlando International Airport. (Photo courtesy of the Orlando Sentinel.)

While the bond sale for the eastern extension was in progress, the FTA was moving ahead with the western project. Consulting engineers had studied four alternative routes before recommending the one which ended at the Turnpike's Orlando-South interchange. It would upgrade the existing McCoy Road to limited access by building parallel service roads from S.R. 15 to a point midway between Daetwyler Road and Orange Avenue. There, the new road would angle southwest, cross Orange Avenue and end at the South Orange Blossom Trail. All traffic from the toll road would get on Orange Blossom Trail briefly before entering the Turnpike to reach I-4. By eliminating an interchange at Orange Avenue where traffic was estimated to be light, Rex was able to offer an interchange at S.R. 436 and the Bee Line, if Orange County could provide the right-of-way. Since a four-laned S.R. 436 was scheduled to open at that point in 1969, the interchange was highly desirable. But that plan was abandoned when, at Governor Kirk's insistence, it was decided to extend the road beyond the Turnpike and connect it directly with I-4. Noting that such an extension would cut 10 miles off the trip from Tampa to the Cape and relieve I-4 congestion downtown, Rex authorized a traffic and earnings study to get it started. But he explained that there was not enough money to build both the extension and the S.R. 436 interchange. For the time being, S.R. 436 and Daetwyler would have to be served by grade crossings.[95]

At a May meeting, the SRD, the FTA, the OOCEA, and the Orange County Commission approved the extension to I-4 and the delay of the interchange. Present at that meeting were representatives of Florida Gas and Martin-Marietta who together donated about $700,000 worth of right-of-way for the extension.[96] Rex planned to build the new extension in three segments with bids being advertised in June, August, and near the end of the year.

Shortly after the May 1969 meeting, the FTA was abolished in a major governmental reorganization following ratification of Florida's new 1968 constitution. Its functions were absorbed by the new FDOT which succeeded the old State Road Department. Instead of the five-member State Road Board which had managed the SRD, the new agency was headed by a single secretary. While the new agency attempted to keep FTA projects moving, it was hampered by a lengthy struggle between Republican Governor Claude Kirk and the Democratic state legislature over appointment of a Secretary of Transportation. The delay unavoidably slowed progress on the Bee Line extensions.

Hopes for an early 1972 opening had faded in May 1971 when Wiley N. Jackson accepted a contract to build the road between the McCoy Jetport and the Turnpike. Work was to begin within 160 days and be completed within 510 calendar days. At the same time, the engineering design work was underway on the Turnpike to I-4 segment with construction contracts still to be let.[97]

The eastern extension was also affected by the governmental reorganization, but there was another problem as well. Environmentalists had raised objections because the Brevard County marshes were the only breeding grounds for the endangered dusky seaside sparrow. Work was halted on that section of the road for 11 months while environmentalists studied its probable impact on the endangered bird. Approval to continue right-of-way acquisition for the road to the Orsino Causeway was not granted until mid-1971.[98]

Construction was soon underway on the disputed section. By December 1971, Wiley Jackson was at work on a 5.3 mile, two-lane road leading to the Orsino Causeway. At the same time,

Construction of the Bee Line extension near McCoy Jetport about 1973.

Looking west across the St. John's River from the fork in the Bee Line which leads to Cocoa and Cape Canaveral to the lower left and to Titusville in the lower right. This eastern extension of the Bee Line was completed in 1974.

earth was being moved on the southern fork of the road. Hubbard Construction Co. was nearing completion of a 5.25 mile stretch from S.R. 520 to the St. John's River.[99]

Except for some material shortages, there were no more major delays on the Bee Line extensions. With six contractors working on the western extension and five others busy on the eastern end, anticipated completion dates varied, but expectations were that the work would be done by late 1973. Total cost of the 11 contracts was about $27 million. When the work was finished, a car could travel from I-4 to the Atlantic Coast for 70 cents. There was a 15 cent toll plaza between McCoy Jetport and the Turnpike, a plaza just east of S.R. 15 where 35 cents would be collected, and another about a mile east of S.R. 520 where the toll was 20 cents.[100]

The four-mile section between McCoy and the Turnpike was opened to traffic in late July 1973, and the response surprised everyone. Announcing a daily average of 2,184 paying vehicles during the first week of operation, R. W. Stevens, FDOT toll facilities bureau chief, said, "We didn't expect that volume…until the link was completed to Interstate-4."[101] The Turnpike to I-4 section opened in December, coinciding with the removal of the last construction barriers on the East-West Expressway. The eastern section opened without fanfare on February 16, 1974.[102]

Central Florida drivers and tourists were enabled to drive on a limited access highway from I-4 to the Atlantic Coast. But there was still considerable work to be done on the road. The two-lane northern fork leading to the Orsino Causeway would eventually have to be widened to four lanes on right-of-way which was already in place. It had been understood that an interchange at S.R. 436 and the Bee Line had only been delayed by lack of funds. The grade crossing at Daetwyler Road was also inadequate. Both would have to be addressed. The need for these two improvements became more acute when the recently created Greater Orlando Aviation Authority (GOAA) began planning for a vastly expanded terminal to handle the burgeoning traffic at its Orlando International Airport.

Looking east from the toll plaza near S.R. 520 across the St. John's River. This is the portion of the Bee Line extension in Orange County which was built by Hubbard Construction.

Looking west across the Bee Line-S.R. 520 intersection from which the eastern extension of the Bee Line began.

With plans for a 1980 opening of the airport, the GOAA asked for assistance. There being no funds for such a project, the OOCEA was selected as the agent to obtain financing and coordinate with other concerned agencies, including the U.S. Navy, the Federal Aviation Authority (FAA), the GOAA, the FDOT, the Orange County Commission, and the city of Orlando.[103]

While Post, Buckley, Schuh & Jernigan, Inc., (PBS&J), a Miami consulting engineering firm with Orlando offices, conducted a preliminary study of the interchange, plans expanded to include improvements to the Bee Line itself along a 2.5 mile corridor extending a half mile east of the interchange and two miles west. The portion from Florida Road (now Tradeport Drive) to McCoy Road would be raised 20 feet to pass over Via Flora and Daetwyler. There would be an interchange at Florida Road. The plan was developed over a three-year period and approved by the OOCEA in August 1979 with some minor adjustments later that year. With inflation increasing costs by about $200,000 a month, estimates ranged from about $14 million in 1978 to about $19 million in 1981.[104]

The necessary right-of-way required negotiations with the FAA to avoid interference with the airport's emergency landing system as well as to coordinate the height of the raised roadway to keep it below the glide path of landing aircraft. That the new right-of-way would cut into a golf course on the Navy base was another problem. Despite loud roars of protest from retirees who used the facility, a cooperative Navy base commander agreed to move the golf course to the rear of the installation and the GOAA agreed to pay for its construction. The GOAA also donated a sizable amount of land for right-of-way and access roads. Private land was also acquired, some of which required condemnation proceedings.[105]

The new toll plaza on the western end of the Bee Line (a Turnpike project) was completed in 1973.

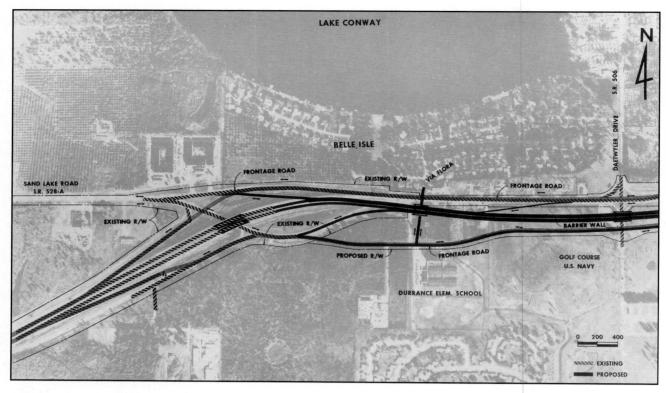

This corridor study of the Bee Line improvement project was completed by PBS&J in 1977. The new road was built almost as shown except that Boggy Creek Road was extended from the south to connect with it.

The largest problem remained the question of how to finance the project. The engineers recommended increases of tolls on the East-West Expressway and the Bee Line to help pay for a bond issue. County Commissioner Lee Chira opposed such a measure on the grounds that those who used the facility should pay for it. He recommended a new toll plaza be constructed at the airport interchange. That was opposed by some who thought charging to get to the airport would leave a poor impression, while others pointed out that it would tie up traffic at what was certain to be a busy intersection. While no decision on Chira's suggestion was made at the time, a study was implemented to help decide whether and where a toll plaza might be located. Chira was speaking for himself, but the entire County Commission opposed a pledge of its secondary gas tax funds, an act which was essential if bonds were to be sold.[106]

After nearly three years of preliminary planning, discussion, and negotiations, both engineering and financing began to take shape in 1979. While the county had urged construction of the interchange only, with the corridor improvements to follow later, the OOCEA and the FDOT favored building both phases at the same time. With financial prospects improving, the OOCEA voted in August 1979 – with minor changes in November – to approve PBS&J's designs and build both phases together. The cost estimate in November was $19.5 million, nearly three times the cost of the original 17.4 miles of the Bee Line.[107]

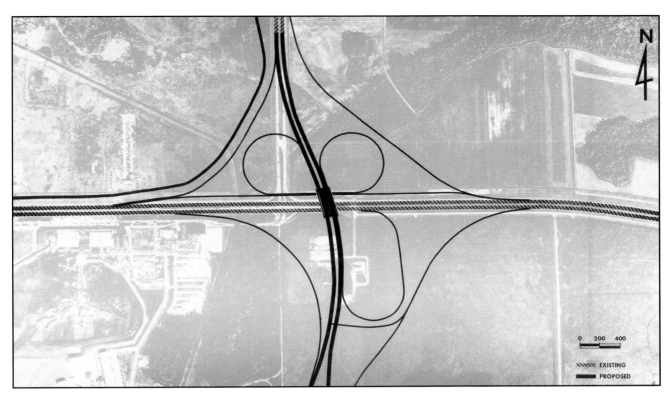

This corridor study of the Orlando International Airport interchange was also completed by PBS&J in 1977. Because of increases in traffic projections before construction began, the final design of the interchange had to be expanded.

On the financial front, the Orange County Commission had changed its position several times, but eventually recommended that the OOCEA look for other sources of funding before coming back to the Commission as a last resort. The county's position was not an unreasonable one. It argued that the 1980 Project (Airport Interchange - Bee Line Improvement) would be beneficial to the state in general as well as the local area. Realizing that state road funds were in short supply, but that the project carried a high priority, the OOCEA authorized Jim Greene to speak to Governor Bob Graham about including the project in a supplemental budget request. The Governor agreed, but the legislature dropped it from the 1979 appropriations measure. "It just fell through the cracks," lamented Senator George Stuart, Jr.[108]

At Governor Graham's urging, FDOT Secretary William Rose managed to find $8.6 million, roughly half of the amount needed, which he committed to the 1980 Project as a grant. An elated Jim Greene said that bonding was a strong possibility for the remainder needed. But there was still a bit of controversy over the FDOT grant. Apparently dissatisfied with the agency's partial funding of a project which had been denied by the legislature, Representative Fred Jones, chairman of the House Transportation Committee, conducted a hearing and requested that the OOCEA be present on February 2, 1980, to discuss the matter. Executive Director John L. Gray, Jr., who had succeeded Harvey Gaines in 1977, attended the hearing as the OOCEA

Governor Bob Graham, 1979-1987, was instrumental in finding FDOT funding for the 1980 Project which enabled the OOCEA to qualify a bond issue for the remaining amount needed.

Executive Director John L. Gray succeeded Harvey Gaines in 1977 and served until 1985.

representative. After expressing his opinion of the matter, Representative Jones asked Gray how the OOCEA had obtained the money. Gray replied, "By check."[109]

The OOCEA prepared a bond issue for the balance of the funds required. It raised tolls at East-West Expressway plazas and ramps by five cents. An increase from 35 to 50 cents was levied at the Bee Line plaza. The increases were to be effective May 1, 1980. With its potential for revenue thus enhanced, the OOCEA went back to the Orange County Commission for its pledge of secondary gas tax funds. One reason for the county's reluctance had been that the OOCEA had been obliged to borrow $5.45 million to cover shortages in the East-West Expressway debt management fund during its first four years of operation. The East-West was self-sufficient by 1979. Perhaps that made the difference. In any event, the Orange County Commission voted in October 1979 to pledge funds from its secondary gas tax revenues to back an OOCEA bond issue of up to $20 million to finance construction of the airport interchange and Bee Line corridor improvements.[110]

While financial negotiations were in progress, accelerating growth in Central Florida had stimulated another FDOT traffic study in the airport area. In January, the FDOT reported estimates far greater than the earlier ones which had been used in designing the interchange. The new figures necessitated additional lanes and other design changes. With the increased projections of traffic and continuing inflation, the FDOT was then estimating costs of the entire project from $27.5 million to $29.9 million. Don Barton, consulting engineer from PBS&J, while agreeing with the FDOT's growth figures, said he did not believe that the cost would be that high. With the new figures in hand, the OOCEA decided to issue $17 million worth of bonds. They were sold in January 1981 at 9.23 percent, a rate in keeping with the time. Barton was correct though. A $20.9 million construction

contract was let that same month to Wiley N. Jackson Construction Co. to complete the project. It was familiar ground to Jackson, who had just recently built the four-lane improvements over much of the same route.[111]

The opening date of the Orlando International Airport had been extended to September 1981, but that still meant that road construction would be competing with traffic to and from the new facility for more than a year. Drivers endured considerable hardship during the overlap, but construction was finally completed in January 1983. Governor Bob Graham attended the dedication ceremony for the Airport Interchange - Bee Line Improvement on January 21, 1983.[112]

As it was finally designed, the airport interchange is shown here under construction in January 1982.

The much discussed new toll plaza, which Commissioner Lee Chira had recommended, was also built, but it was located on the Bee Line just west of the airport interchange. It opened on July 2, 1983, with a toll of 25 cents for all vehicles.[113]

Long before the Airport Interchange - Bee Line Improvement project was completed, the OOCEA was already addressing a lengthy list of other additions to the expressway system, including prospects for a beltway, or beltline, as it was then called.

Construction of the realigned Bee Line where it curves to the southwest from near McCoy Road. (Photo courtesy of the Orlando Sentinel.)

Building A Community

Construction in progress near the old Jetport terminal in January 1982.

OOCEA Chairman Jim Greene is shown here at a dinner following dedication of the Airport Interchange- Bee Line Improvement project in January 1983 recognizing Governor Bob Graham for his effort in making the project possible.

Orange County Commission Chairman Lou Treadway accepting an award recognizing the county's support of the project.

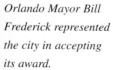

Orlando Mayor Bill Frederick represented the city in accepting its award.

Richard Fletcher, Harvey Gaines, and Phil Reece conversing after the program ended.

Chapter 5

Central Florida's Beltway Vision

First proposed as the "Outer Belt Route" in 1950, a bypass around Orlando was envisioned as a 35-mile roadway whose cost was estimated at about $4 million. By 1965, when a State Road Department study showed a need for an "outer beltline," the city had expanded, costs had risen, the required road had lengthened, and the means for financing it were just as elusive as ever. Although the Orlando Urban Area Transportation Study (OUATS) included an outer beltline in 1965, no specific route had been selected and no agency had been designated to build it. The Orange County Commission had contracted with Campbell, Foxworth and Pugh, and VTN, Inc., to study a western beltline starting at I-4 in the area of the proposed Maitland Interchange and running southwest to reconnect with I-4 in the Disney World area. There was a general understanding that an eastern beltline should connect with it on the north side and proceed southeast, crossing the East-West Expressway and the Bee Line before entering northern Osceola County toward an intersection with I-4.

In 1971, with the East-West Expressway moving rapidly toward the construction phase, OOCEA Chairman James Greene asked for and received FDOT permission to look into the feasibility of constructing a beltline around Orlando.[114] He soon found that financing was not the only obstacle.

In early 1972, the Orlando, Seminole, Osceola Transportation Authority (OSOTA) was organized to take over and operate a bus system which it hoped to expand. It also had authority to build roads, but was not empowered to condemn land for right-of-way or to sell bonds.[115] Several community leaders suggested OSOTA be given these additional powers so that it could build the outer beltline. At the same time, the FDOT favored the OOCEA as the builder because it had a record of successful building and could better market its bonds. A series of discussions involving the Orange, Seminole, and Osceola County Commissions, the OOCEA, the OSOTA, and the FDOT, resulted in a tentative agreement whereby Orange County would continue its preliminary work with the western beltline, while the OOCEA would begin working on an

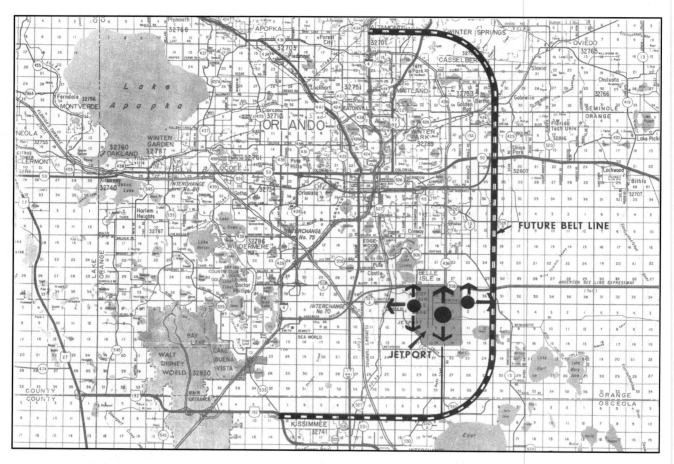

The eastern beltline as it was envisioned in the early 1970s.

eastern road which it would build pursuant to an interlocal agreement, still to be worked out, between it, the three counties, and OSOTA.[116]

Proceeding on this informal arrangement, Greene and the OOCEA requested a loan of $150,000 from the FDOT for preliminary financial and engineering studies toward a proposed $100-$150 million eastern leg of the beltline. In October, the OOCEA approved HNTB to do the preliminary engineering work and Wilbur Smith and Associates to study traffic and earnings.[117]

In the meantime, Greene had continued negotiations on the interlocal agreement, a draft of which called for a five member authority including the chairmen of the Orange, Seminole, and Osceola County Commissions and the directors of the OOCEA and OSOTA. After more negotiations and several amendments, all parties except the FDOT had signed the agreement by January 1973, but the Orange County Commission had added a proviso that it could back out if all parties had not signed the agreement within 60 days.[118]

Just as the agreement was being signed, the confrontation referred to in Chapter 3 between Building Movers, Inc., and the HNTB engineers occurred. Building Movers had run several months behind schedule in removing buildings and Diversacon, Inc., was being delayed in beginning construction. A confrontation between Building Movers personnel and HNTB engineers Harry Bertossa, Dale Patten, and John Kelly had been resolved acrimoniously in late December, two months after HNTB had been selected to perform preliminary financial and engineering work on the Outer Beltline. Coincidentally, the terms of Richard Fletcher and E. C. Goldman expired in January 1973, and they were shortly replaced by Richard Swann and William Poorbaugh. With unseemly haste, a meeting was called, and Greene was immediately ousted from the chairmanship to be replaced by Swann. The first order of business was to reopen the search for an engineering firm to conduct preliminary work on the Outer Beltline. Calling it "Operation Incredible," the Orlando Sentinel declared the change a ploy "aimed at 'firing'" HNTB from the beltway project.[119] It was at this point that Governor Reuben Askew intervened, demanding that OOCEA members return Greene to the chairmanship or resign. Poorbaugh resigned, but the others voted – some of them reluctantly – to restore Greene to the office. That ended two weeks of chaos and J. Thomas Gurney was subsequently appointed to replace Poorbaugh.[120]

But the damage was done. Citing the FDOT's failure to sign the agreement within the required 60 days and the apparent inability of the OOCEA to manage its affairs without gubernatorial interference, the counties withdrew their support.[121]

The project languished until September 1973 when FDOT Secretary Walter Revell approved a joint venture with the OOCEA to go ahead with the feasibility studies. Acting as agent of the FDOT, the OOCEA contracted with HNTB and Wilbur Smith and Associates to study a possible route covering about 33.5 miles from the Florida Turnpike in Osceola County, through eastern Orange County, and extending to I-4 in Seminole County. The two firms were to report their findings in June 1974, by which time, of course, the oil embargo and resulting gasoline crisis were having their adverse effects on automobile transportation. In the interim, Seminole County organized its own expressway authority, at least partially to have a voice in decisions relating to the tollway within its boundaries.[122]

The results of the engineering studies were disappointing. Harry Bertossa of HNTB said a tollway would have only a "marginal" chance of paying for itself. In their presentation to the OOCEA, Bertossa, and Norman Wuestefeld of Wilbur Smith and Associates, showed that neither of the two routes selected – one costing $144 million and the other estimated at $140 million – would produce enough revenue to amortize the debt at prevailing interest rates. Sam Draper, bond development coordinator for the FDOT, confirmed the engineers' assessment. In further discussion, it was agreed that the failure of the East-West Expressway to meet projections since its recent opening was a contributing factor to the gloomy reports.[123]

OOCEA Executive Director Harvey Gaines concluded that it was too early for the beltline, but time would cure that. Speaking of the fine line between too few automobiles to support a toll road and too much growth driving up right-of-way costs, Jim Greene speculated that "there's no good time to build a toll facility. You're either ahead of time or behind time."[124]

By the late 1970s, "behind time" seemed to be gaining ground. Traffic was increasing and toll revenues were rising on both the Bee Line and the East-West Expressway. The OOCEA was working as fast as possible to complete the Airport Interchange - Bee Line Improvement (the 1980 Project) in anticipation of a deluge of traffic from a much-enlarged airport. Demographers were predicting more growth, more tourists, and more traffic.

Improved planning since the early 1960s gave metropolitan Orlando leaders a much better grasp of transportation needs than had been possible in earlier years. A 1962 federal law had required state and local governments to develop systematic plans for transportation needs to make their projects eligible for federal funding after 1965. While the legislation did not directly effect the OOCEA because toll roads were ineligible for federal funding, it had the

The History of the Orlando-Orange County Expressway Authority

indirect effect of demonstrating to local government leaders the essential role the agency could play in filling the gap between the roads needed in metropolitan Orlando and the funds which were available to build them.

From this federal requirement and the readily observable mismatch between automobile traffic and the roads available for it, the Metropolitan Planning Organization (MPO) evolved in 1977. With representation from the three counties, most of the municipalities, the FDOT, the OOCEA, and even the Orange County School Board, the MPO became the overall transportation planning agency for the area. With no implementation powers, it maintains the official transportation plan for the area and annually prepares a transportation improvement program to guide the implementing agencies. By 1979 it was projecting the need for a half billion dollars worth of road construction by the year 2000, an amount far beyond the anticipated revenue available for it.[125]

The Orlando Urban Area Transportation Study Year 2000 Needs Plan.

The MPO plan clearly demonstrated a need for the services which the OOCEA could provide. It specifically included, for example, three large toll projects. The Airport Expressway, later referred to as the Central Connector, was a 5.5 mile road intended to connect downtown Orlando with the airport by way of the Bee Line. It called for an extension of the East-West Expressway to connect with a Western Bypass. And it also included an Eastern Bypass,

Building A Community

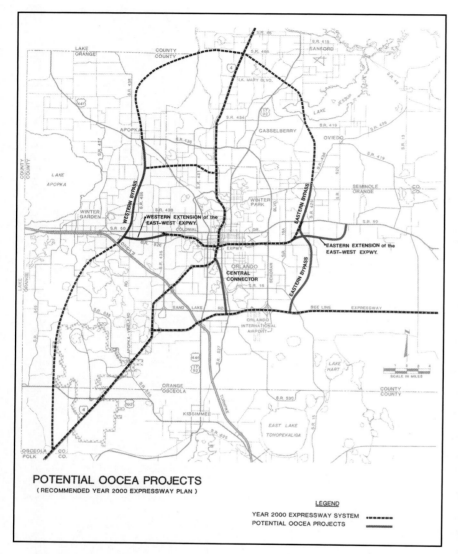

Potential OOCEA projects recommended by PBS&J in the Long Range Expressway Plan.

envisioned as an alternate to S.R. 436, to run from the East-West Expressway to the Bee Line.[126] Even Seminole County leaders were seeking cooperation with the OOCEA on the Eastern Beltway by the early 1980s.

The 1983 Long Range Expressway Plan

With its role as a contributor to the metropolitan transportation system seemingly accepted by other agencies, and with revenue from tolls on the East-West Expressway and the Bee Line increasing at a steady rate, the OOCEA was enabled to take a broader and longer view of its activities. It commissioned PBS&J, its general engineering consultant, to study the metropolitan area's need for expressways by the year 2000 and recommend suitable toll road projects which the OOCEA might undertake to contribute toward filling that need. In the meantime, the OOCEA began planning for improvements of the East-West Expressway to accommodate Orlando's remarkable growth. At the request of the Orange County Commission, which advanced funds for the project, it also agreed to study an extension of the East-West Expressway and determine a center line for it.[127] With an $800,000 loan from the FDOT, it was concomitantly doing preliminary engineering work on an Eastern Beltway to run from the Bee Line to the Seminole County line.

As the OOCEA's consultant, PBS&J had broad responsibilities beyond the engineering and financial planning included in the Long Range Expressway Plan. It was also charged with general program management, coordination with consultants, environmental permitting, and the design of lighting, signalization, and toll facilities. The toll facilities design required close coordination with the FDOT and collection equipment vendors.[128]

Finished in 1983, the Long Range Expressway Plan identified six projects which seemed suitable for construction by the OOCEA. They were each part of a whole expressway system, but could be individually viable roads. Ranked number one in priority was the four mile northern section of the Eastern Bypass (from S.R. 50 north to S.R. 426). The reason for the ranking was the rapid growth in the area. Unless right-of-way was secured soon, costs would force its relocation eastward. Second ranking was given to a five mile eastern extension of the East-West Expressway, where rapidly rising land prices also necessitated early reservation of right-of-way. Third in priority was the southern section of the Eastern Bypass, 7.6 miles between the East-West Expressway and the Bee Line, so located as to provide opportunity for eventually extending the beltline south around the airport.[129]

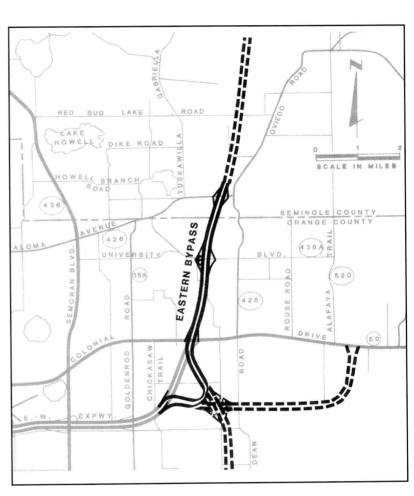

The Eastern Bypass from S.R. 50 to S.R. 426 was recommended as Priority One.

Ranking fourth was the mid-section of the Western Bypass (ultimately to become Western Beltway) combined with the Western Extension of the East-West Expressway. The mid-section of the Western Bypass (all within Orange County) was a 9.2 mile road from the Florida Turnpike to U.S. 441 east

The Eastern Extension of the East-West Expressway was Priority Two.

of Apopka. It was a classic example of Greene's earlier lament about timing. Growth was driving up land prices and making routing difficult, but its length made it extremely costly for the time being. The accompanying Western Extension was a four mile road intended to connect the East-West Expressway with the Western Bypass. Ranked lowest in priority was the Central Connector (Airport Expressway). The five-mile road connecting downtown with the Bee Line was expected to be extremely costly (estimated at $100.7 million in 1983 dollars) because of the intensive development through which it would have to be built.[130] The six enumerated roads were all deemed to be feasible, but some would need additional revenues in the form of grants, donated right-of-way, and possible toll increases.[131]

According to PBS&J, the third priority was the southern portion of the Eastern Bypass connecting the East-West Expressway with the Bee Line.

The History of the Orlando-Orange County Expressway Authority

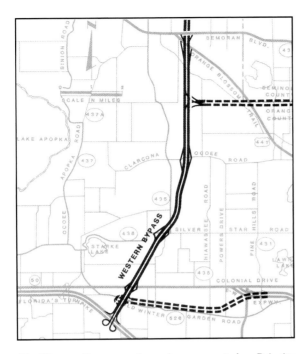

The Western Bypass as shown here was rated as Priority Four. In subsequent planning it was moved westward.

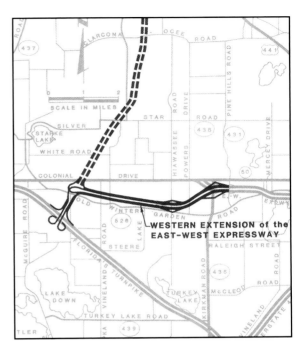

The Western Extension of the East-West Expressway was viewed as part of the Western Bypass. But when that road was realigned to the west, the Western Extension was redesigned to connect directly with the Florida Turnpike.

Envisioned as a rapid route from downtown Orlando to the new airport, the Central Connector was recommended to run from I-4 in the downtown area to the Bee Line west of the airport. Its high cost rather than its need was the reason for its sixth rating.

Improved Financial Capability through Debt Refunding

A major refunding which greatly improved the OOCEA's financial capabilities was accomplished in 1985. The bonds issued in 1965, 1970, and 1980 carried covenants requiring that all toll revenues be used exclusively to retire the debt. Revenue was increasing enough by the mid-1980s to create a surplus over what was needed to manage that debt and pay other expenses. If that surplus were freed from the restrictive covenant, it could be used to improve the OOCEA's financial standing and enable it to sell bonds for other projects. Deciding to pursue that course, it contracted in 1985 with M. G. Lewis and Company to refund its debt with terms permitting it to use its surplus for new construction. By purchasing some of the outstanding bonds on the open market the OOCEA was able to refund $80 million in existing debt with a new issue of $57,715,000. That issue was sold in December 1985 at 8.25 percent. The result was a savings of $27 million over 25 years and a greatly increased bonding capacity for new construction. A new lease purchase agreement was completed with the FDOT by which the state agency agreed to continue paying for maintenance of the original Bee Line, as well as both operations and maintenance on the East-West Expressway, and the Airport Interchange - Bee Line Improvements.[132]

There was strong approval when the OOCEA announced that it was preparing a $433 million bond issue which would enable it to complete most of its ambitious building program between 1988 and 1991, about five years earlier than previously anticipated, as well as to spend about $10 million on improving the existing East-West Expressway.[133] Approved by the Governor and Cabinet in April 1986, the bond sale was held up by two law suits brought by landowners who were disgruntled over the routing of two of the proposed new roads. But the suits were thrown out and the bond sale was completed in November 1986. With the sale, the OOCEA became one of the largest road builders in the state. Its lease purchase agreement with the FDOT on this project differed from earlier ones. The OOCEA was obliged to pay for operating and maintenance costs on newly constructed roads while the FDOT guaranteed only to make the payments if the OOCEA was unable to do so.[134]

In early 1987, a contract was signed with Public Financial Management to manage OOCEA funds so as to produce the most revenue while maintaining a cash flow that would pay its obligations.[135]

In 1989, to handle the OOCEA's growing financial program, Gregory Dailer was employed as director of finance.

Less popular than the new building program was one of the measures undertaken to make the bond sale possible. With advice from Vollmer and Associates, its traffic and earnings consultant, the OOCEA announced plans to double most tolls on the East-West Expressway and the Bee Line in January 1987, and then increase them again by an equal amount in 1990. The announcement brought extensive complaints, but OOCEA officials explained that it was the only way to build the roads which most people agreed were necessary. An extensive publicity campaign, proclaiming that "The Expressway is the Best Way," accompanied the January 1, 1987, implementation of the first increase. While the number of cars on the East-West dropped for a while, there was little change on the Bee Line. Even though traffic dropped, revenue rose dramatically. Within a year, most people had adjusted. Revenue of about $14 million in 1985 rose to about $22 million in 1987 and $28 million in 1988. Most people seemed to have forgotten that a comparable increase was pending for 1990.[136]

As part of an advertising program to explain the toll increases, this brochure outlined the four parts of the 1986 Project which would add 26 miles of new roads to the expressway system.

Improvements of East-West Expressway Ramps and Toll Plazas

The growing volume of traffic had overtaxed the ramps and toll plazas on the East-West Expressway by the mid-1980s. Cars were backing up at both ends of the road, at several ramps, and at the toll plazas. The funds provided in the 1985 bond issue for this purpose were already being put to use. Two lanes had been added to the Holland East Toll Plaza in 1984 and it was soon being expanded to 14 lanes, while the Holland West Toll Plaza was being widened to 12 lanes. Both had reversible lanes for the rush hours. Lanes were also being added at S.R. 50 on both ends of the expressway as well as at critical ramps such as Conway Road and Rosalind Avenue. These were in place by 1987.

Toll Cheaters and Toll Gates

Since the opening of the East-West Expressway, toll managers had to deal with a few robberies and considerable vandalism. FDOT maintenance logs showed that collection baskets had been clogged with hot chocolate, honey, glue, paper clips, paper, razor blades, and bubble gum. But the greatest and most costly problem was the number of people who simply drove through without paying. The honor system had not worked as well as hoped – and expected. The FDOT had some success with the employment of off-duty policemen to watch the collection baskets and arrest violators. A fine of $48.50 could be exacted for infractions. But that was an expensive remedy. Gerry Coleman, a toll supervisor, summed it up succinctly, "You're not going to catch them if you are not there. If you're there, they don't do it." There was more to it than that, though. At one ramp where a policeman was present, 14 persons in one hour not only did not pay, but had no money on them when they were caught.[137]

When a 1985 study concluded that toll cheaters amounted to more than five percent of the motorists using the toll road at an annual revenue loss of $500,000, the OOCEA decided it was time to install toll gates. First installed in February 1986 at the Holland East Toll Plaza where losses were greatest, the gates reduced losses by more than $300,000. Gates were then installed at the Holland West Plaza and some ramps. A considerable improvement in toll collections, the gate system was not entirely satisfactory. Costing $48 each annually, the gates sometimes failed to operate, they were easily broken, and people felt that they slowed traffic. But they reduced the violation rate to less than three percent and the OOCEA decided to use them until a better solution was found.[138]

Financed by proceeds from the 1988 bond sale, collection was further improved in 1989 by the addition of manned facilities at Orange Blossom Trail, Mills, Bumby, and S.R. 436.[139]

The History of the Orlando-Orange County Expressway Authority

Personnel Changes and Additions

Beginning with the surprise resignation of Chairman Jim Greene in January 1985, there were several changes on the OOCEA board during the 1980s. Coinciding with the extensive building program then underway, they were accompanied by a number of additions - and one change - to the staff.

Greene's resignation was quite a surprise to his colleagues and the community at large. Although some of his friends thought he might have been exhausted from the many controversies surrounding plans for building the northern portion of the Eastern Beltway, most noted that he had been volunteering his services to the community for 14 years and that it was time for him to concentrate on his personal business and his family.[140]

George M. Barley, Sr., who had served as OOCEA treasurer since 1975, was ill and would also leave the board within a year. Wilbur S. Gary had only been on the board since 1983. That left Wayne P. (Phil) Reece as the only healthy member with lengthy experience on the board. Like Barley, a member since 1975, Reece had been an active board member and was knowledgeable of the massive building program then being planned. To fill Greene's vacancy, the Governor appointed Allen Arthur, another capable person who had twice served on the board in the late 1970s and early 1980s as a representative of the Orange County Commission. He had knowledge of the OOCEA's building plans, and his appointment was well-received. In 1985, Tom Dorman was the ex officio member from the Orange County Commission. A harmonious board easily elected Phil Reece to succeed Greene.[141]

Reece served successfully for two years before resigning to become a member of the Florida Transportation Commission, a newly formed panel which would set policy for the FDOT. When George Barley resigned in late 1986 for health reasons, and Allen Arthur's term expired shortly afterward, Betty Jean (B. J.) West and Robert S. (Bob) Harrell were appointed to fill the vacancies. Jacob Stuart succeeded Reece.[142]

Wayne P. "Phil" Reece was a member of the OOCEA from 1975 to 1987 and chairman during the last three years. Reece, assisted by OOCEA attorney J. Fenimore Cooper, and Keith Denton of Paine Webber, worked out the refinancing of the old debt which enabled the agency to finance the 1986 Project. The planning and design of that project were also completed during his chairmanship.

Robert S. Harrell, OOCEA member from 1987 to 1991 and chairman 1987-1988 and 1990-1991.

Betty Jean (B. J.) West was chairman of the OOCEA from 1989 to 1990.

Then, shortly after Bob Harrell was elected chairman, new legislation required that the district director of the FDOT should serve as an ex officio member of the OOCEA. A brief round of resignations and new appointments left a new board composed of three members appointed by the Governor, the incumbent chairman of the Orange County Commission, and the regional director of the FDOT. They were Bob Harrell, chairman; B. J. West, vice chairman; Wilbur Gary, treasurer; Lou Treadway, chairman of the Orange County Commission, ex officio; and Ben Watts, District FDOT director, ex officio. Jacob Stuart had resigned so that Harrell could continue as chairman. Harrell then resigned the chairmanship in early 1989 and was succeeded by B. J. West.[143] Executive Director John Gray announced his resignation to become effective in early 1985. David W. (Bill) Gwynn, then director of research and development for the New Jersey Department of Transportation, was named to succeed Gray. He took over the position in April 1985.[144]

In early 1985, the OOCEA's only employee was Susan Simon, the office manager, who had assumed that position in 1978, replacing Myrtle Rizza who had held it since 1965. Simon was joined by Darleen Mazzillo in 1986 and Patricia Varela and Vicki Coleman in 1987.

J. Fenimore Cooper continued as OOCEA legal counsel, a position in which he had served with only a brief interruption since 1964.

Because of the resignations of Greene and Gray, and the burgeoning building program then in progress, there were several additions to the professional staff in the following months.

In January 1985, while Gray was still on the job, Charles C. Sylvester, Jr. was employed on contract to act as liaison between the OOCEA, the FDOT, and PBS&J. Sylvester was well suited to the position. Recently retired from the FDOT's district office in DeLand where he had served

David W. (Bill) Gwynn was executive director of the OOCEA from 1985 until 1991. Because of his professional qualifications and the expanding scope of the OOCEA's activities, Gwynn was delegated more authority than his successors, and his management of the 1986 project was a major accomplishment.

as its liaison with the OOCEA, he had been actively involved in the agency's planning and construction since the early 1970s. He was thus able to provide continuity in the building program while Gray was being succeeded by Gwynn. Beyond that, a civil engineer well-informed about Central Florida land use for many years, he became the OOCEA's right-of-way expert during its ambitious building program. He remained in the position until 1998.[145]

In the meantime, Joseph McNamara became the OOCEA's first assistant director. He was succeeded in 1988 by Joe Berenis. That same year, Jewel Symmes joined the staff as manager of communications.[146]

To handle the impending building program, two other new positions were created. William (Bill) McKelvy became director of construction in 1986, first by contract and, after 1991, as a regular employee. Also a longtime FDOT engineer who had worked with Sylvester out of the DeLand office before retiring in 1983, he was closely acquainted with Central Florida road building since first working on the I-4 project. About the same time, John Dittmeier became director of planning for a brief period.

Susan Simon (right), an OOCEA employee since 1978, and Ingrid Nielsen, who worked with her for a brief period, are shown here speaking with Governor Bob Graham when he spent one of his "work days" with the agency.

Building A Community

After acting as FDOT liaison officer to the OOCEA since the early 1970s, Charles C. Sylvester, Jr. then became a consultant with the agency from 1985 to 1998.

The Beginning of a Beltway

Plans for a full beltway were still incomplete in the mid-1980s, but 35 years after it was first discussed and nearly 15 years after Jim Greene had proposed its initial construction to the FDOT, its first two segments were in the advanced planning stage as part of the $433 million dollar 1986 Project.

Joseph A. Berenis has worked with the OOCEA since 1988 as director of engineering and, since 1992, as deputy director of the agency.

William B. McKelvy, who has worked on the Bee Line and many other Central Florida projects as an FDOT engineer, became director of construction for the OOCEA in 1986 and retired as director of construction and maintenance in 2001. He was replaced by Ben Dreiling.

Chapter 6

The Beltway's Beginning Segment:
The Northeastern Beltway

Of the six major projects PBS&J recommended in the 1983 Long Range Expressway Plan as appropriate for construction by the OOCEA, it named the Northeastern Beltway as its first priority for two reasons. First, that road promised to carry a larger traffic load from the outset, thereby enhancing the agency's financial ability to proceed with other projects. Second, in 1983 there was a "generally vacant corridor" available for a suitable route which would have little impact on existing developments. But anticipated new development would likely soon close the available options if action was not taken quickly to reserve the necessary right-of-way. Deciding to move ahead as quickly as possible, the OOCEA agreed that the Northeastern Beltway segment would be its first priority.[147]

In the fall of 1983, PBS&J recommended for the Northeastern Beltway a route running six miles – the existing portion from Goldenrod Road to S.R. 50 was to be improved – from the East-West Expressway generally west of Dean Road and connecting with S.R. 426 in Seminole County. The county approved the general route, a step essential to the project because the OOCEA needed its continuing pledge of gas tax funds to support its bonding capacity. Since a limited access road in the area was part of the MPO's current transportation plan, and it was consistent with Orange County's Growth Management Plan, the OOCEA, meeting in January 1984, declared the road a necessary part of the urban area's transportation system by the year 2000.[148] But there were many obstacles in the way before construction could begin.

First, it was determined to keep the public informed about plans and how they would affect residents and property owners. Second, legislation had been enacted creating a Department of Environmental Regulation (DER) which was empowered to grant or deny permits for construction that might have adverse affects on wetlands, water recharge areas, and other sensitive natural areas. Permits were also necessary from the St. John's River Water Management District.

Third, it was necessary to coordinate its activities with the Seminole County Expressway Authority (SCEA). That agency, envisioning an extension of the expressway through its county, had hired Gerald Brinton as executive director and Wilbur Smith and Associates as general engineering consultant. A Project Development and Environmental Study (PD&E) was completed and a final alignment of an 18-mile road through the county was adopted in 1987. At that point negotiations were opened between the OOCEA and the SCEA to coordinate connection between the Northeastern Beltway and the Seminole County Expressway. Their objective was an interlocal agreement about who would build the Seminole County portion. The already tenuous negotiations were made even more so by actions in the state legislature.[149]

There was considerable discussion in Central Florida about the need for a building agency with multi-county authority. A proposed Metropolitan Transportation Authority, supported by city leaders, had failed to win voter approval in 1985. It was followed in 1987 by an effort on the part of Orange and Seminole County legislators to create a Regional Expressway Authority with power to build in the two counties. It too failed, but, as will be shown, it did exacerbate the difficult task of negotiating an interlocal agreement to build the small portion of the Northeastern Beltway in Seminole County.

Early in the project, PBS&J notified people who lived or owned land along the proposed route and held several public information meetings. Thus interest was high when another session was scheduled at Valencia Community College on January 25, 1984. About 600 people attended the meeting and 35 of them voiced their concerns. It was clear that many were opposed to any road which would take their property or create traffic noise near it. Several developers were also concerned that the road would interfere with their own plans for building and selling homes in the area. As planned at that time, the Dean Road route would have affected about 60 houses and two businesses.[150]

Opposition was already organized. The University Boulevard Coalition, representing 14 homeowners associations and led by Martin Goodman and Judge Michael Cycmanick, organized about five years earlier in opposition to

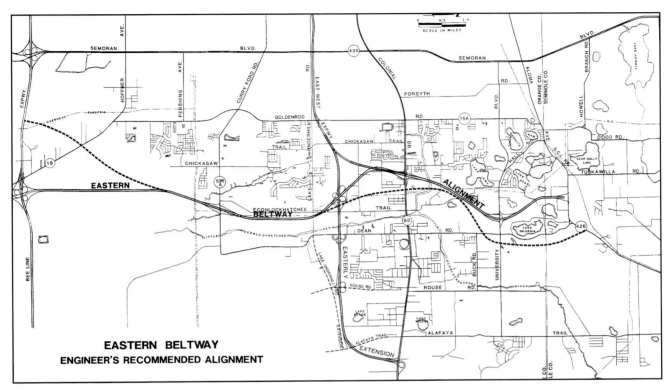

Engineer's recommended alignment for the Northeastern Beltway. The Dean Road alignment (shown in two solid parallel lines) was the preferred route. The alternate alignment (shown in dashed line) in the University Boulevard-Lake Georgia area was preferred by the University Boulevard Coalition.

commercial development on that road, turned its attention to the Northeastern Beltway. Letters and telephone calls by the hundreds poured into the offices of the county commissioners. Numerous citizens attended OOCEA meetings. Public interest was so great that the OOCEA scheduled its regular February 22 meeting at the Bob Carr Performing Arts Center to make room for the crowd. A day earlier, citing the outpouring of protests, the County Commission had voted unanimously to ask the OOCEA to consider a more easterly route.[151]

The OOCEA agreed to a study of two additional routes, one near Rouse Road (east of Dean) and the other east of the University of Central Florida. Completed in early April, the study affirmed the advantages of the Dean Road route over the other two. After a week of intense negotiations, the Orange County Commission reversed itself and agreed to let the OOCEA decide the route of the Northeastern Beltway. The agency then told PBS&J to develop a center line for the Dean Road route.[152]

Having lost the political battle, the University Boulevard Coalition became a defender of the environment. The PBS&J plan called for an interchange at University Boulevard which would destroy part of a 25-acre wetland area. Determined to continue its opposition by concentrating on the regulatory front, the Coalition hired Irby Pugh, an environmental attorney, as well as David and Casey Gluckman, a husband and wife team considered to be the most influential environmental lobbyists in Tallahassee. It then demanded an environmental study of the land in question. The study, estimated to take about three months, would further delay progress on the Northeastern Beltway.[153]

The Coalition had about $19,000 in its account to pay legal expenses. About $10,000 of that amount had come from two developers whose land was in the path of the new road. According to the Orlando Sentinel, Ludwig Goetz, owner of the Orange Ridge property, had contributed $6,000. The paper also reported that another $4,000 had come from Robert Mandell of Greater Construction, whose firm was then developing Watermill. There were five other subdivisions in the path of the new road in various stages of development. Some, including Mandell, already had county approval of their plans and were building. "We are in the business of selling houses," Robert Mandell told a Sentinel reporter, "There's no sense to stopping the development now."[154] Others were trying to obtain approval during the delays. Ludwig Goetz, admitting that he was doing it only to increase the value of his land, submitted plans for his Orange Ridge subdivision. At the request of the OOCEA, the county denied his application. Goetz then filed suit against the county. The suit was ended when the OOCEA agreed to pay Goetz about $1,400,000 for 12 acres of right-of-way.[155]

The OOCEA approved a legal description of the center line of the Northeastern Beltway in August 1984, and in September the Orange County Planning and Zoning Commission recommended that the county reserve a 300-foot-wide right-of-way for the Dean Road route. In early 1985 the OOCEA applied to the DER for a permit to dredge and fill at several points along the Northeastern Beltway proposed route. The most critical area involved the 25 acres on the south side and five acres on the north side of University Boulevard where the interchange was to be located.[156] The University Boulevard Coalition continued its campaign while the OOCEA waited for the DER to rule. It distributed thousands of cards for its members to mail to various governmental leaders, and held demonstrations at the contested site. In November 1985, concerned about the loss of "some very good wetlands," the DER disapproved the OOCEA's plans, but said a solution might be worked out. With the University Boulevard Coalition and several environmental groups continuing to object to the routing of the road, a compromise was eventually arranged. In exchange for destroying 12.8 acres, the OOCEA agreed to create 13 acres of new wetlands, create a storm water retention area along a drainage canal flowing into the Econlockhatchee River, and turn over to the state about 19 acres of wetlands south of University Boulevard to prevent their development. Stephen Fox of the DER said this "mitigation" would not harm water in the area and could produce a "slight benefit." Martin Goodman, presumably speaking for the University Boulevard Coalition, said, "It's my intention to continue the fight."[157]

Despite the opposition to the Dean Road route and the complications arising from negotiations with other agencies, progress was being made. Having undertaken a building program involving four major projects and then accelerating its scheduled completion date by nearly a decade, the OOCEA made operational adjustments to handle the large workload and keep its schedule. One overall result was the delegation of authority to the executive director to do many of the things which Richard Fletcher and James Greene had done in the past. With his qualifications for the job and with a staff to assist him, Bill Gwynn was given broad responsibilities to act subject to board approval. For example, with dozens of contractors working on lengthy projects, a consent agenda was adopted whereby the 30-40 monthly invoices could be approved in a single motion (subject to individual exceptions) without taking up the time better used for policy decisions.

Gwynn was empowered to carry out those policy decisions, coordinate matters with other agencies, and represent the OOCEA before the public, subject to the oversight of the board. An example was the OOCEA's delegation of authority to Gwynn to make contractual changes necessary for the expeditious completion of construction contracts in amounts not to exceed 15 percent of the base contract amount.[158] In a related effort to speed up construction, a modular building was located near the OOCEA administrative offices to house the construction management engineering firm (Zipperly Hardage and Associates) which had overall supervision of construction.[159]

The net effect of the vastly increased building program, the employment of an experienced highway engineer as executive director and a qualified staff to assist him, and the willingness of the OOCEA chairman and his colleagues to delegate adequate authority to him, was the professionalization of the OOCEA. In the past Richard Fletcher and James Greene had dedicated enormous amounts of their time to carry out exceedingly important duties simply as a public service to their community. In addition to making policy, they had carried out most of the day-to-day activities necessary to implementing it. Phil Reece, who had also given freely of his own time for more than a decade, saw the necessity of employing a professional executive director and a staff to

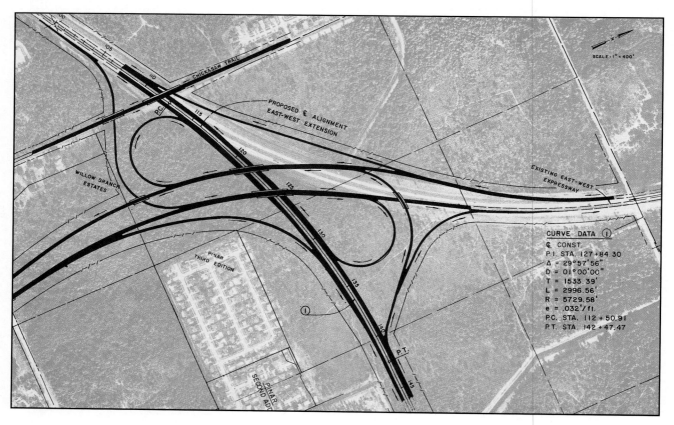

Proposed design of the East-West Expressway-Eastern Beltway (408/417 interchange) beginning just west of Chickasaw Trail (top left). Note the old East-West Expressway to S.R. 50 paralleling the new Eastern Beltway (center right).

support him in carrying out a mission which was no longer manageable by one person. In making this change, Reece and his colleagues performed a major service to the OOCEA and the community it serves.

In October 1984, the firm of Gee and Jenson contracted to prepare preliminary engineering plans for the Northeastern Beltway. Working closely with PBS&J, the firm completed plans, including interchange configurations, in August. It also furnished aerial photographs to be used for the final design and cost estimates for right-of-way construction.[160]

In December 1985, a contract was made with Greiner Engineering to complete final designs for Section I of the Northeastern Beltway (from Goldenrod Road to a point just north of S.R. 50, including a flyover for Chickasaw Trail and interchanges at Valencia College Lane and S.R. 50). Two contracts for that service on Section II were made at the same time with Dyer, Riddle, Mills and Precourt. The first was for the road from north of S.R. 50 to the county line, including an interchange at University Boulevard. The second was for the portion in Seminole County, including a grade connection with S.R. 426 and the widening of that road from Dean Road to Tuskawilla Road. One of its responsibilities according to the latter contract was to make arrangements with the SCEA regarding that portion of the road.[161]

With Charles Sylvester maintaining general supervision over right-of-way acquisition, Harvey Gaines was once again employed by contract to manage it. The OOCEA employed several appraisers from firms in the area on a fee basis to appraise the property.[162] Appraisals began as soon as right-of-way maps were available in July 1986. Offers were made to owners on the basis of the appraisals – moving costs were given consideration at this point. If the offer was not accepted – and many of them were not – J. Fenimore Cooper would then initiate eminent domain proceedings. In that case, the property would be taken and the amount offered placed in escrow. A jury would decide the final amount.

There were slightly more than 200 parcels in the path of the Northeastern Beltway, with about 92 houses and four businesses. Many of the houses were

This house was given a temporary reprieve when it was made the field office of the Hardaway Construction Co.

House at 9700 Trevarthon Road awaiting removal while the construction contractor was already clearing right-of-way for the Northeastern Beltway.

individual residences, but some were owned by developers who had continued to build as long as possible. Laurel Homes, Inc., was paid $2,228,500 for its property taken for right-of-way. The Greater Construction Company received almost $2,400,000. As previously mentioned, Ludwig Goetz was paid about $1,400,000. By far the largest sum was paid to Malcolm Clayton, Charles Clayton and E. G. Banks. For a 53-acre tract – needed for the Northeastern Beltway as well as the Eastern Extension of the East-West Expressway and a large interchange – and several other properties, they were paid more than $11,000,000. The total cost of right-of-way for the Northeastern Beltway was about $35,000,000.[163]

The prime contractors cleared the sites. Hubbard Construction Company, for example, ended up with 31 vacant houses which were sold in a block to Modern Moving and Wrecking Company. That firm moved them to other sites for buyers who eagerly purchased them.[164]

While right-of-way acquisition was proceeding, the OOCEA prepared for construction. According to Bill McKelvy, director of construction, with three other projects following the Northeastern Beltway in quick succession, the OOCEA was creating a great deal of work for the heavy construction industry and realized that suppliers would be hard-pressed to keep up with demand. It also wished to open its bidding to as many firms as possible, some of which were smaller than others. For these reasons, it decided that the work should be "strung out" in parcels small enough to permit more firms to participate and to stagger the starting dates to even out the demand for supplies.

Accordingly, the Northeastern Beltway was divided into projects 101 through 106 with the latter pertaining to toll plazas.[165]

One other time-saving measure was suggested by McKelvy and the

Looking east across the Northeastern Beltway as construction begins at Trevarthon Road. The Little Econlockhatchee tributary (E-4 Canal) is on the right.

construction management firm. Because the lead time was so great in the delivery of structural steel and reinforced earthwall panels, they recommended that these materials be purchased in large quantities to be fabricated and stored by the manufacturers for delivery to the sites as needed. Gwynn was authorized to make the purchase for the Northeastern Beltway and the impending Eastern Extension of the East-West Expressway in amounts not to exceed $3,000,000 for structural steel and $2,000,000 for reinforced earthwall panels.[166]

In January 1987, the firm of Zipperly Hardage and Associates (ZHA) was selected as construction manager for both the Northeastern Beltway and the Eastern Extension. The $2,900,000 contract was for a period of 25 months. ZHA occupied offices in the modular building near the OOCEA offices where it maintained close contact with Director of Construction Bill McKelvy and Executive Director Bill Gwynn as well as the several construction firms in the field.[167]

Advertisements for bids for construction, engineering and inspection (CEI) contractors went out in April. These contractors were responsible for ensuring that the prime contractors met acceptable standards in soil testing, compaction of earthen banks, placing of structural steel and reinforced walls, application of concrete on road surfaces and other matters of quality control. Parsons Brinkerhoff won the contract for those services on Section I of the Northeastern Beltway (from Goldenrod to north of S.R. 50, including projects 101, 102, and 301). The contract was for $1.9 million for a 20 month period. Section II (from north of S.R. 50 to S.R. 426, including projects 103, 104, and 105) was contracted to Reynolds, Smith & Hills for the same amount and time.

Governor Bob Martinez, 1987-1991. (Photograph courtesy of Florida State Archives.)

However, all costs relating to project 105 (about $300,000) were to be paid by the SCEA. The CEI contract for the toll plazas went to Mizo-Hill, Inc.[168]

Actually the first project relating to the Northeastern Beltway was not among those numbered 101 through 106. It was the construction of ramps at Goldenrod Road to complete the diamond interchange, the western half of which had been built with the original East-West Expressway, and the closing of Bryan Road. That work was contracted to Martin Paving Company of Daytona Beach for $1,158,824.82. With 270 calendar days for the entire project, that contract was awarded on the same day that the invitation to bid went out for projects 101 through 105.[169]

Ground-breaking ceremonies for the Northeastern Beltway were held on July 15, 1987. A marching band played while Governor Bob Martinez stepped out of his helicopter to make a brief speech. A number of pickets were there and one woman demanded that Martinez comment on the appraisal of her house, with which she was dissatisfied. The Governor politely declined, saying that he was not an appraiser. Many in the crowd were happy about the new road, but obviously not everyone approved.[170]

The contract for Project 101 (from Goldenrod Road to north of Valencia College Lane, including the resurfacing of the old road, a flyover at Chickasaw Trail and an interchange at Valencia College Lane) went to K & L Contractors for $7,947,775.14. Work began in July, shortly after the opening ceremony. Chickasaw Trail was to be closed for about a year and a temporary crossover was installed at Valencia College Lane.[171]

The prime contract for Project 102 (north of Valencia College Lane beyond S.R. 50 to the Little Econlockhatchee, including

Clearing right-of-way was easier in some places than others. This backhoe was buried in muck near the tributary of the Econlockhatchee River.

Pilings are in place and bridge work is beginning at the Little Econlockhatchee tributary. In the center left equipment can be seen at work on a bridge which will cross the beltway.

resurfacing of the existing road to S.R. 50, went to Martin Paving Company for $9,391,359.79. A notice to proceed was issued in late July and the company began work in early August 1987 relocating utilities and extending box culverts which had been installed on the old East-West Expressway, and removing buildings from the right-of-way.[172]

Hardaway Construction Company was the prime contractor for Project 103 (from the Little Econlockhatchee to just south of University Boulevard). The contract for $5,839,032.71 was signed on July 30, but work was delayed briefly while a right of entry was being obtained from one property owner.[173]

A contract to build Project 104 (from south of University Boulevard to the county line) was awarded to Hubbard Construction Company in late August in the amount of $6,682,622.[174]

Project 105 was the 2,000 feet of road from the Seminole County line to S.R. 426 and the widening of that road from Dean Road west to Tuskawilla Road. Negotiations had been underway since late 1986 on an interlocal agreement whereby the OOCEA would act as agent for the SCEA in building the road. Executive Director Bill Gwynn and engineers from Dyer, Riddle, Mills and Precourt had met repeatedly with SCEA personnel and an agreement was finally completed and approved by the OOCEA in late July 1987. Bids for the project were opened on September 9, and the OOCEA met on September 30 to award the contract. Just before that date, the SCEA requested that the agreement be amended to say that, if a new Regional Expressway Authority was formed, all funds held in escrow for the project would be returned to Seminole County and the SCEA would be reimbursed for any costs it had incurred during the previous 60 days. The SCEA had tabled the agreement until

Concrete being poured for the deck of the northbound bridge over the Econlockhatchee.

Having served its purpose, the field office became the last structure removed from the right-of-way. It was demolished on December 15, 1988, before the new road opened the following day.

its next meeting on November 18. Gwynn was not pleased. In order to keep to its tight construction schedule, the OOCEA was compelled to issue the contract that day (September 30). He recommended, and the OOCEA board agreed, that "we turn over all of our work product and plans to the Seminole County Authority and let them handle the project as their own."[175]

The SCEA, having preferred all along to build the road, was not at all displeased by the OOCEA decision. In 1986, it had arranged for a $5,000,000 line of credit backed by pledges of future toll receipts and a portion of the county's gas tax revenue.[176] Some of that money had already been used to purchase right-of-way. Using about $4,100,000 of the remaining amount, the SCEA quickly advertised for bids and let a contract for construction of the project as it was already designed. The work was completed on time and that segment of road opened in late December 1988 along with the Orange County portion of the Northeastern Beltway.[177]

Project 106 was for construction of the toll plazas. The contract was awarded to Southland Construction whose low bid was $3,030,754.05.[178]

Construction proceeded without major difficulties. There were some complaints about road closing, smoke from burning debris, and the moving of several million

Looking north along completed Northeastern Beltway. S.R. 50 is in the center.

yards of dirt by eleven hauling companies. But despite complaints of dust, noise, and traffic congestion, the dirt moving problem was nothing like the Forsyth Road situation 15 years earlier.[179]

About two weeks ahead of schedule, the Northeastern Beltway was opened on December 16, 1988. Governor Bob Martinez, DOT Secretary Kaye Henderson, and OOCEA Chairman Phil Reece participated in the opening ceremony. Vollmer and Associates had projected about 12,000 vehicles a day when the road first opened. During the first few days, which were toll free, about 17,900 cars used the road each day. There was a drop after tolls were implemented on January 1, 1989, but in March of that year, 1,176,176 drivers paid $329,355.78 to use the road. The six mile road cost about $105,000,000, about $35,000,000 of which was for right-of-way.[180]

Meanwhile, construction of the Eastern Extension of the East-West Expressway was moving rapidly toward completion.

Building A Community

Governor Martinez addressing a crowd at the dedication of the Northeastern Beltway on December 16, 1988.

OOCEA employee Darleen Mazzillo (far right) with ZHA employees at the dedication ceremony.

Phil Reece speaking with Governor Martinez after the ceremony.

Less than half a mile inside Seminole County near S.R. 426, this sign marked the northern end of the Eastern Beltway until the Seminole One project was completed.

Chapter 7

The Eastern Extension of the East-West Expressway

The Eastern Extension was ranked second in priority in the 1983 Long Range Expressway Plan. Expected to carry 30,000 vehicles daily by the year 2000 as a result of anticipated growth in east Orange County, its high priority was the result of a high revenue-to-debt ratio and the related fact that rapid development was threatening the selected alignment. The OOCEA accepted the premise that it should move as rapidly as possible to reserve right-of-way for that alignment.[181]

The Orange County Commission also favored this road for early consideration, and advanced $157,000 for a study by Gee and Jenson, a traffic engineering firm, to determine a center line. The county was interested because Martin-Marietta was planning to open a plant on Lake Underhill Road in 1984 and the Huckleberry Planned Unit Development of some 6,877 units was anticipated nearby. Orange County was planning a southward extension of Alafaya Trail and an eastward extension of Lake Underhill to serve these and other developments.[182]

A public information meeting in January 1983 had alerted landowners and residents along the corridor that the road was being planned. A second meeting was held in April at the East Orange Community Center at which the Gee and Jenson study was presented. Several hundred residents attended. It was explained that the study had been undertaken to find a route which avoided as many houses, mucklands, and lakes as possible. But there were numerous objections. Martin-Marietta favored extending the expressway, but objected to the route presented because it would "completely landlock" the Martin land. Huckleberry opposed it because there was no interchange planned for Alafaya Trail (not yet built). High Point of Orlando, another planned unit development, for which an interchange was envisioned, complained that the road would isolate part of its property, "drastically reducing its value." A large number of homeowners were fearful of being forced to sell their homes and move. OOCEA Executive Director John Gray explained that finding a suitable strip of land was "like threading a needle" because it was flanked on both sides by subdivisions and planned developments.

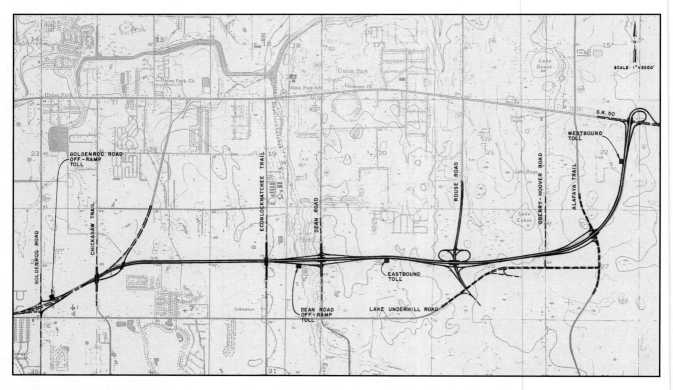

Corridor location study of the Eastern Extension of the East-West Expressway.

Once the OOCEA accepted a legally described center line for the road, Orange County had authority to reserve the land for that use, with the understanding that the OOCEA would purchase it within five years. The comments at the January meeting convinced the OOCEA officials that a better understanding needed to be reached before it asked the county to reserve a specific route. Gee and Jenson was instructed to study the concerns of the landowners in the corridor to see what could be done. Its engineers met with Huckleberry, the University of Central Florida, the Central Florida Research Park, John Pepin, a Memphis State University professor who owned a 138-acre tract in the corridor, High Point, and several others, and heard their comments. An adjusted center line legal description was then presented and the OOCEA accepted it on July 25, 1984, but added that the center line "is subject to amendment." The county was asked to reserve the land for the Eastern Extension.[183]

There were reasons for the difficulty in setting a workable center line. The portion from Chickasaw Trail eastward past Dean Road had been designed to do the least damage possible to the homes along the way. Beyond that point eastward, there were too many variables. Not only did the large landowners have conflicting goals, but the governmental agencies also had differing opinions. The original 900 foot corridor had shown a road crossing Alafaya Trail – which had not been built – and curving northward about 2,000 feet to intersect with S.R. 50. Both Orange County and the FDOT preferred that the road be terminated at Alafaya Trail, which would be perhaps six lanes wide to handle traffic between the Eastern Extension and S.R. 50. But the consulting engineers argued that the purpose of the road was to relieve traffic on S.R. 50 and, therefore, a connection with it was essential. Vollmer and Associates and others pointed out that such a connection would improve the traffic count and therefore enhance revenue.[184]

Huckleberry's purposes were best served by the Orange County-FDOT preference because it wanted to develop its land which was directly in the path of the Eastern Extension on the east side of the planned Alafaya Trail. But, the Central Florida Research Park was about to expand southward and wanted a connection with the Eastern Extension, preferably farther eastward on S.R. 50 than the 900 foot corridor would permit. It was willing to donate land if the OOCEA would build the road so that the Research Park could connect to it on the north side of S.R. 50.

Al Benton and Tom Russo of Gee and Jenson, Timothy Jackson of PBS&J, Charles Sylvester, and other OOCEA personnel held many meetings with the various principles and eventually reached compromises that satisfied most of them. It was agreed that the road would cross Alafaya Trail and curve gradually northward, intersecting S.R. 50 nearly a mile to the east. That made the Eastern Extension a six-mile road. The interchange would allow a connection with the Central Florida Research Park, which donated some right-of-way.[185]

That route was made possible by months of negotiations with the Huckleberry officials who finally agreed to sell 79 acres – at a price of $1,460,294 – for right-of-way and to reserve other land for ten years for a potential extension of the road even farther eastward.[186]

The center line was shifted slightly southward to the advantage of the High Point Planned Development. High Point donated some right-of-way and the OOCEA agreed to build two large drainage culverts under the new road to facilitate drainage. John Pepin was more persistent. Both parties agreed that his property would be damaged by the road, but they differed sharply on the remedy.

There were many meetings with Pepin, but the OOCEA considered his demands unreasonable. He had once asked for three bridges over his land and $700,000. When that failed, he filed suit in the summer of 1986, attempting to block the entire OOCEA building program. Fearing that the suit would prevent the impending sale of bonds, J. Fenimore Cooper exclaimed, "This is hurdle number two thousand." But in September the Florida Supreme Court dismissed Pepin's case. The OOCEA was thus enabled to sell its bonds and keep its 1986 Project on schedule. But Pepin was not finished. He filed another suit in Orange County Circuit Court in the fall of 1987, saying that the Eastern Extension would landlock his property. After several months the suit was settled. The OOCEA got the 13 acres it needed, and the recalcitrant marketing professor got a bridge across his property – at an estimated cost of more than $2 million dollars.[187]

Despite the lengthy negotiations and a number of verbal agreements, actual right-of-way acquisition did not begin until all the legal steps had been taken and right-of-way maps were available. The appraisers were consequently obliged to move rapidly if the project was to be kept on schedule. Harvey Gaines handled the acquisitions as he had for the Northeastern Beltway. A staff of appraisers, working on a fee basis, systematically appraised each of the parcels which were to be taken all or in part. If the offers based on the appraisals were not accepted by the owners, condemnation proceedings were implemented whereby the court decided the amounts to be paid. A comparatively large number of owners along the Eastern Extension went to court, and some whose residences were being taken, were still unhappy with the compensation they received. An estimated 37 homeowners were obliged to sell and about a third of them went to court. Ten families, whose homes on Deerwood Avenue in the Bay Run Subdivision were in the path of the main exchange, all chose the legal path. With the beginning of construction pending, their cases were decided and all were ordered to leave in February 1987. In several other condemnation cases, owners left, the amount offered was placed in escrow, and the court decided the appropriate compensation later.[188]

About 150 parcels were affected by the Eastern Extension, including those from which only a portion was taken. According to the Orlando Sentinel, the right-of-way for the road cost about $32 million.[189]

Shortly after Gee and Jenson completed preliminary engineering plans for the Eastern Extension, contracts were let for final designs. Sverdrup & Parcel Associates, Inc., was chosen for that work on Phase I from the existing S.R. 408 near Chickasaw Trail to a point east of Dean Road, including the main interchange which connected the northeast, east and southeast legs of the expressway, an interchange at Dean Road, and flyovers connecting with the existing road. The contract amount was not to exceed $1,050,000. Phase II, from east of Dean Road to S.R. 50, including interchanges at Rouse Road, Alafaya Trail, and S.R. 50 was awarded to Prime Design, Inc., for $895,398.[190]

Looking west across the main 408/417 interchange under construction in April 1989. The Chickasaw Trail flyover is in the upper left.

As mentioned earlier, Zipperly Hardage and Associates was named construction manager for the project in January 1987. In July 1987, bids were accepted for CEI services. Kaiser Engineering was assigned to Projects 302 and 303 at a sum not to exceed $1,900,000. DeLeuw Cather Engineering was to be paid no more than $1,700,000 for its services on 304 and 305. The

Looking east across the main 408/417 interchange along the Eastern Extension. The Econlockhatchee Trail is in the upper center and Valencia College is in the center left.

contract for CEI services for Project 306, the toll plazas, went to Mizo, Hill and Glass in the amount of $980,000.[191]

By late September 1987, while work was just getting underway on the Northeastern Beltway, prime contractors were being selected for the Eastern Extension. The contract for Project 301, the road from the existing East-West Expressway to the main interchange, was awarded to Hewitt Contractors which made the low bid of $4,085,716.82.[192]

Project 302 was the main interchange connecting all the roads. That project was contracted to the Rogers Group Contractors for the sum of $9,712,877.52.[193]

Project 303 included the segment from just east of the main interchange to just east of Dean Road, with a full interchange at that road. Hubbard Construction Company received the contract for $7,329,557.[194]

Project 304 was a stretch of road extending from east of Dean Road to east of Alafaya Trail, including interchanges at both. Hubbard was also the low bidder for the project and agreed to complete it for $11,820,255.05.[195]

Project 305 included the much-negotiated portion from east of Alafaya Trail to S.R. 50, with an interchange at the latter road which included access from the Central Florida Research Park. The prime contractor was the Rogers Group which submitted the low bid of $6,972,423.[196]

As with the Northeastern Beltway, Southland Construction Company was awarded the contract (Project 306) for the toll plazas.[197]

After the many obstacles encountered in preparing for construction, the actual building of the road was comparatively without incident. OOCEA officials, the construction managers, the prime contractors, and the suppliers kept to a tight schedule and remained on time. Except for complaints about several roads being closed to make way for the expressway, and too much dust and noise from the heavy dirt trucks, there was only one major problem. An unexpected lawsuit was filed over land adjacent to the right-of-way from which Hubbard Construction had planned to take 620,000 cubic yards of dirt and move it with off-road dirt movers. Deprived of that convenient source, the firm had to obtain the dirt from a more distant borrow pit and move it by truck. That caused an increase of dirt trucks on the roads as well as a supplemental agreement allowing Hubbard an additional $1,500,000 for embankment material.[198]

The first part of the road from Goldenrod Road to Rouse Road opened on May 12. At the dedication ceremony, in anticipation of other projects then underway, OOCEA Chairman B. J. West told the crowd that, "We hope we bore you to death with these openings."

The second portion opened in June, keeping the entire project on time. The overall cost was $105,000,000 with right-of-way accounting for about 30 percent of it.[199]

Looking west from the mainline toll plaza toward Dean Road on May 25, 1989.

Looking west from Econlockhatchee Trail across the completed main 408/417 interchange toward downtown Orlando.

Governor Martinez and entourage arriving by helicopter at the dedication of the Eastern Extension on May 12, 1989.

Governor Martinez addressing the audience at the dedication of the Eastern Extension.

B. J. West, OOCEA chairman, at the podium at the dedication of the Eastern Extension.

Chapter 8

The Southeastern Beltway: *East-West Expressway to the Bee Line*

The Southeastern Beltway, an estimated 7.6 miles of road connecting the Bee Line and the East-West Expressway, was ranked third in priority in the 1983 Long Range Expressway Plan, because it was anticipated to carry 29,000 vehicles daily by the year 2000 and would provide opportunity to extend the Eastern Beltway southward around the International Airport through an area where future growth was expected. Several land owners were already making plans for large residential and commercial developments in that region, and the implementation of those plans was contingent upon an adequate system of roads.[200]

The OOCEA's general consultant, PBS&J, began studying a corridor for the proposed road in 1983. Unlike the other three portions of the 1986 Project, the Southeastern Beltway was planned for a largely undeveloped area where few residences would be affected. But that did not mean that selecting a route would be easy. In fact, a few early public meetings resulted in a comparatively amicable agreement with the residents just south of Lake Underhill Road where most of the houses were. The engineers were able to route the road between the Florida Power Company property on the west and Riverwood Village on the east without taking any homes. Riverwood Village residents were concerned about traffic noise, but they were assured that trees would be planted as a buffer between their property and the highway. Beyond that, a right-of-way southward toward Curry Ford Road was agreed upon and a center line was accepted in July 1984.[201]

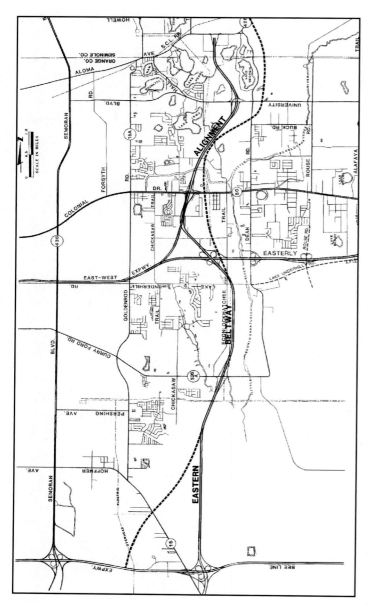

Engineer's recommended alignment for Southeastern Beltway.

The greatest difficulties arose over the mostly vacant land between Curry Ford Road and the Bee Line. Not only did the planners have to keep in mind the environmental impact of the route they selected, but several owners of large tracts were vying to have the road located where it would enhance the value of their respective parcels. At the same time, the Greater Orlando Aviation Authority (GOAA) favored a route as close as possible to Orlando International Airport.

When the original center line was made public, two groups wanted changes. Three land owners – Howard Scharlin, Tony Robaina (both Miami developers) and John Brunetti (owner of the Hialeah Race Track) – were willing to donate one and a half miles of right-of-way, if the center line were moved eastward through their land. Attorney Hal Kantor, speaking for the Lee Vista project (the T. G. Lee property), said that, in return for a shift to the west, his client would donate right-of-way, build feeder roads (Lee Vista Boulevard) and make other concessions. He added that the growing number of people working there needed a road to the north and would add measurably to the traffic on the proposed toll road. In view of the two requests to shift the road in different directions, the OOCEA tabled the matter of the southern section for 30 days.[202]

The delay was much longer. Both groups continued to press their case before the OOCEA and in consultation with Timothy Jackson of PBS&J. Another complication arose when the GOAA asked that a decision be delayed until it had its revised master plan completed in May 1985. The OOCEA agreed, but the plan was not in hand until early 1986. By that time the PBS&J consultants were studying four different routes, each starting from a common point north of Curry Ford Road. They were also preparing "in-depth" environmental assessments of the four alignments.²⁰³ At the July 1986 OOCEA meeting, Bill Gwynn reported that his staff and PBS&J consultants recommended the eastern alignment of the proposed road, because it would cost approximately $15 million less than the others. Anticipated revenues would be about the same, and it would be compatible with future plans to extend the road to the south. The board accepted the recommendation and further authorized PBS&J and Hal Scott – whom the OOCEA had employed as an environmental consultant – in conjunction with the DER and the St. John's River Water Management District, to do an assessment to decide what had to be done to make the proposed road compatible with the environment. About 19 acres of wetlands were affected. Land for the proposed right-of-way was reserved in April 1987.²⁰⁴

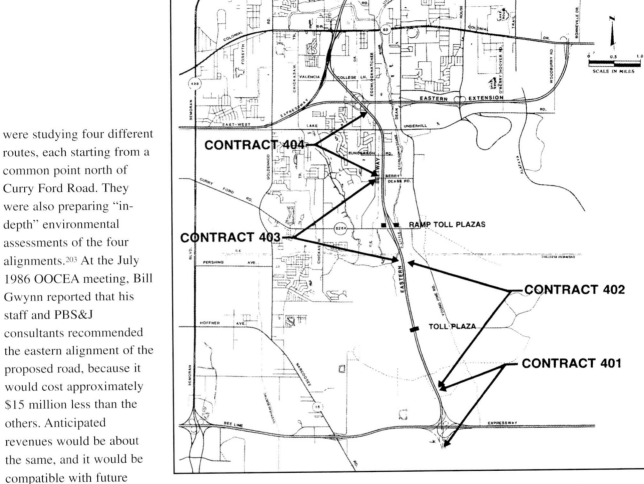

Final design of the Southeastern Beltway.

As had been the case with the two preceding projects (Northeastern Beltway and the Eastern Extension) the OOCEA had to provide new wetlands and maintain them for five years to mitigate the environmental damages resulting from the road construction.

Looking north from the Bee Line along the route selected for the Southeastern Beltway before the heavy construction began.

After advertising for bids and committee review of the 22 proposals received, the OOCEA selected two firms in November 1986 to complete the final design of the Southeastern Beltway. Dyer, Riddle, Mills & Precourt was chosen for the section from Lake Underhill to south of Curry Ford Road, and Prime Design, Inc., was selected for the portion from that point south to the Bee Line. The contracts amounted to $863,000 and $900,000 respectively.[205]

With final designs nearing completion in early 1988 and construction expected to begin in the fall, the OOCEA began selecting engineering firms to manage construction of the road. After reviewing several proposals from firms wishing to become general construction program manager, the OOCEA selected ZHA whose work on the Northeastern Beltway and the Eastern Extension had been excellent. Their new contract was for an amount not to exceed $3,200,000 and was to take effect June 1, 1988. It included both the Southeastern Beltway and the Western Extension of the East-West Expressway.[206]

Although extensive negotiations with landowners had been underway for some time and would continue into the early weeks of construction, right-of-way acquisition began after the final designs were completed in early 1988. Charles Sylvester once again had general oversight of the process, while Harvey Gaines and a crew of appraisers did the field work, making appraisals, settling the unchallenged cases, and turning over to J. Fenimore Cooper those requiring eminent domain proceedings. There were about 60 parcels affected by the Southeastern Beltway and only about seven houses were taken. About a fourth of the owners went to court, including a majority of those whose homes were involved.[207]

By the time right-of-way acquisition began, most of the earlier offers to donate land had disappeared for one reason or another. Disappointed by the route which was finally selected, John Brunetti complained that the OOCEA was "steamrolling us."[208]

Scorning an offer of $1,700,000 for the requisite right-of-way through his swampy land, he offered to accept $12,000,000 for it. Wishing to settle the matter as quickly as possible, the OOCEA made a counter offer of $4,000,000. Apparently insulted, Brunetti went to court. After a lengthy and complicated trial, he was awarded $4,000,000 (including all costs) for the contested tract.[209]

With the road alignment adjoining their property, Howard Scharlin, Tony Robiana, and Ralph Fisch, as trustee for other owners, wanted an interchange to which Lee Vista Road could be connected. Needing right-of-way for its road, the OOCEA worked out an arrangement whereby it paid about $1,100,000 for right-of-way through four pieces of property and agreed to construct an overpass at an appropriate place for the extended Lee Vista Road. The property owners agreed to extend the road and pay for the ramps at the interchange. The developers first planned to pay for the ramps through a special assessment district tax, but the OOCEA could not accept that method, since it might adversely affect the credit worthiness of its bonds. OOCEA officials asked that the construction be paid for in cash or a certified line of credit prior to November 1, 1988, the deadline for awarding the construction contract. No action was taken in the allotted time.

The overpass was constructed, but neither the extension of Lee Vista Road nor the ramps were built at that time. The project would be revived in the late 1990s.[210]

Looking north from near Berry Dease Road. Lake Underhill Drive is in upper center. Construction toward the main intersection crosses Econlockhatchee Trail in upper left.

Mainline toll plaza between the Bee Line and Curry Ford Road in late December 1989.

Fifteen proposals for CEI services on the Southeastern Beltway and the Western Extension were reviewed by a selection committee in April 1988. Contracts were awarded in August to Wilbur Smith and Associates for Projects 401 and 402 in the amount of $1,700,000, and to Parsons Brinckerhoff for Projects 403 and 404 for $1,950,000. A contract for CEI services for the toll plazas (Project 406) was awarded to GAI Consultants-Southeast, Inc., in June 1989. GAI was to be paid $521,026.[211]

The first construction contracts were awarded in late November 1988. Contract 403, for construction from the Curry Ford interchange to north of Berry Dease Road, went to Hewitt Construction Company for $7,195,293.82. Contract 404, for the road beginning north of Berry Dease Road and ending just south of the centerpiece interchange, was awarded to Sloan Construction in the amount of $7,175,464.82. Notices to proceed on both projects were issued in early January 1989.[212]

The contract for Projects 401 and 402 was delayed because of negotiations relating to the Lee Vista Road interchange. In late December, the selection committee recommended that the Hardaway Company be awarded a contract for both projects (from the Bee Line to the Curry Ford Interchange). The OOCEA agreed to issue a conditional contract for $23,226,846.58 (if the Lee Vista Road ramps were included). If no agreement was reached on Lee Vista, then the contract amount was $20,179,754.85. A notice to proceed was issued on January 11, 1989.[213]

Project 406, for construction of the toll plazas, went to Southland Construction Company for $5,054,124.79 in March 1989.[214]

Probably because construction was beginning later than planned, the starting dates of construction were not staggered, but efforts to avoid delays by

The Bee Line and Eastern Beltway interchange under construction in March 1990.

prepurchasing supplies continued. For both the Southeastern Beltway and the Western Extension, $2,370,855.85 worth of prestressed concrete girders and piling was purchased from Durastress, Inc., and Reinforced Earth, Inc., was paid $1,564,401.57 for precast wall panels.[215] Construction proceeded without interruption and even the weather cooperated. When the first section of the road, from the East-West Expressway to Curry Ford Road, opened ahead of schedule on April 14, 1990, Bill McKelvy explained that, "Weather was a major factor…If you're growing corn the weather is too dry, but if you're building roads it's just right."[216] The section from Curry Ford Road to the Bee Line opened on June 26, still ahead of schedule, but just in time for the toll increase scheduled for July 1, 1990.[217]

As finally designed and built, the Southeastern Beltway was nine miles long. It cost $72,000,000 with right-of-way representing only about $13,000,000 of it.[218]

Meanwhile the Western Extension of the East-West Expressway to the Florida Turnpike was also nearing completion.

Brochure explaining the reasons for and benefits to be expected from the 1990 toll increase.

Looking south to the Bee Line-Eastern Beltway interchange, completed and open for traffic.

Only four minutes away, the new Eastern Beltway was expected to be beneficial to Orlando International Airport.

Chapter 9

The Western Extension of the East-West Expressway: *The Turnpike Connector*

In the 1983 Long Range Expressway Plan, the Western Extension had been recommended to connect with the middle section of the Western Beltway, but the route of that road had subsequently been moved westward. With that change, a Western Extension connecting with Florida's Turnpike became more useful for the movement of metropolitan Orlando traffic. Recognizing that such a route would require cooperation with other agencies, OOCEA Chairman Phil Reece spoke to the Florida secretary of transportation in early 1985 about it. "All we've done is start a dialogue," he told the Orlando Sentinel, but that dialogue resulted in the possibility of joint funding for the road.[219]

Professional Engineering Consultants W. K. Daugherty Inc. was awarded a $369,907.05 contract in August 1985 for a preliminary engineering study of a possible route. Work on the project began in October 1985.[220]

Although the road would not reach Lake County, residents of that county as well as those in western Orange County, applauded the proposed road because of its improvement of transportation to Orlando from those areas. Lake County leaders also saw it as a boon to growth in the southern part of their county.[221] At a public hearing on the Western Extension at the Orange County Commission chambers in September 1985, it became clear that residents along the proposed route were not as enthusiastic about the road. While many thought the road would be beneficial for traffic, they also wanted to know more about what the road would do to their homes and neighborhoods.[222] Alex Hull and other PEC personnel kept the public informed at meetings in January and again in March 1986 as they developed their preliminary plans.

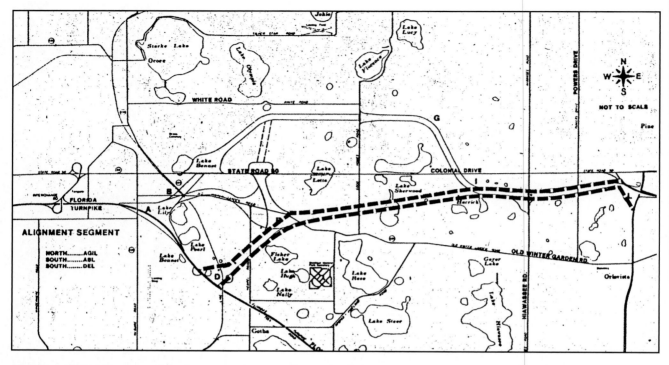

The three alignments proposed for the Western Extension of the East-West Expressway to the Florida Turnpike. The parallel dashes show the route recommended by the engineers.

At a meeting at the Tabernacle Baptist Church in late January, they reported three potential routes, all of which shared the same path west from Kirkman Road past Hiawassee Road. Route One would then loop north of S.R. 50, cross back south, and connect with the Turnpike near Lake Lily. The most expensive of three routes, it would take 211 homes, but it would also carry more traffic and have less impact on the environment. Routes Two and Three would both stay south of S.R. 50 and cross Lake Sherwood, differing only in that Route Three would intersect the Turnpike farther south near Lake Bonnett. At the same time, Hull was meeting with DER personnel about the environmental aspects of crossing Lake Sherwood.[223]

Alex Hull recommended Route Three because it was less costly and disruptive, and the crossing of Lake Sherwood was environmentally feasible. Although it would cut through the Orlo Vista community and cross the lake, it would save at least $5,000,000 and require the removal of fewer homes. Another meeting at the Tabernacle Baptist Church in early April drew a crowd of 350 people, but according to the Orlando Sentinel, "relatively little flak." There was both criticism and approval of the proposed road. But the most persistent critic had already spoken. Brantley Slaughter, president of the Livingston Meadows Homeowner's Association, began his long-running battle with the OOCEA in mid-March, decrying the Western Extension because it

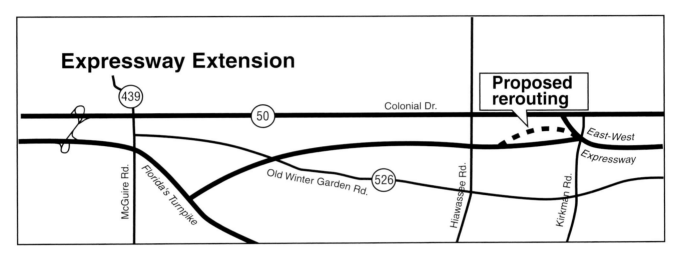

Because the recommended route would destroy so many homes, the engineers found a way to adjust the route as shown here. The adjustment saved 48 homes.

would serve regional needs at the expense of neighborhoods. His recommendation of a double-decked S.R. 50 between Kirkman and Hiawassee Roads as an alternative apparently met with little support.[224] The OOCEA approved the corridor in mid-April and Hull went to work on an alignment.[225]

Tom Kelly of PEC recommended a center line alignment which the OOCEA adopted on July 23, 1986. OOCEA Director of Planning John Dittmier reported that, instead of the county handling the right-of-way reservation process as in the past, it would henceforth be done by the OOCEA pursuant to recent legislation. Yet another meeting was held on July 31 to inform the public that right-of-way was being reserved according to the recently adopted center line. Many residents voiced their concerns that so many homes would be taken between Kirkman and Hiawassee Roads.[226] After further study, the engineers found a way to shift the road a few hundred feet north, so as to reduce the number of homes to be taken. The route approved in July would have taken 128 homes, 122 of them between the two roads. The adjustment reduced that number to 76 with 70 of them between Kirkman and Hiawassee.[227] Yet another public meeting was scheduled for September 30 in the city of Orlando council chambers to discuss the adjusted center line.

President of the Livingston Homeowner's Association, Brantly Slaughter, was still dissatisfied with the route of the Western Extension which would take

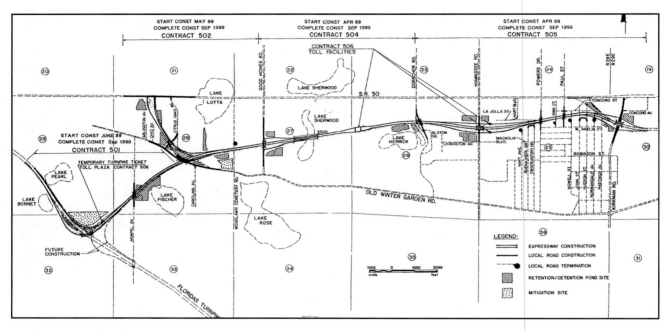

Final design of the Western Extension including access to S.R. 50 via Good Homes Road.

Bridge construction across Lake Sherwood.

part of his backyard. He filed suit against OOCEA members Tom Dorman, Allen Arthur, George M. Barley, Sr., Wilbur Gary, Phil Reece, and the Florida Division of Bond Finance, arguing that the OOCEA's plan to sell bonds to build toll roads was unconstitutional and a violation of his civil rights. The suit was found to be without merit and did little to prevent the bond sale or the construction of the Western Extension of the East-West Expressway.[228]

In November, 22 proposals were received for final design services on the Western Extension and the Southeastern Beltway. Professional Engineering Consultants W. K. Daugherty, Inc. was selected for that part of the road from Kirkman Road to Good Homes Road. The portion from Good Homes Road to the Florida Turnpike was to be handled by Beiswenger, Hoch & Associates, Inc.

Contracts were signed in February 1987 for $1,350,000 and $2,200,000 respectively.[229]

One of the problems to be addressed was the crossing of Lake Sherwood. It was ultimately decided that a 1,000 foot bridge across the lake was preferable to filling part of it for causeways. Coordination was also required with the FDOT regarding the Turnpike connection. An interlocal agreement was completed whereby the OOCEA was authorized to act as agent for the FDOT for construction of the East-West interchange with the Florida Turnpike. The Turnpike District designated $18,000,000 for the design, right-of-way acquisition, and construction of the interchange and the approach to it (Project 501).[230]

Right-of-way acquisition began in mid-1988 with the same personnel applying the same procedures as in the three previous projects, except that the OOCEA rather than the prime contractors handled the removal of houses and other structures. According to the OOCEA's parcel status reports of November 3, 1988, all or portions of about 200 parcels were taken. That included 76 houses, four businesses, one church, and a few other structures.

The approximate cost of the right-of-way was $31,500,000. A fairly large number of owners, especially those whose dwellings were being taken, went through eminent domain proceedings. Brantley Slaughter and some of his neighbors continued to protest, but most of those affected accepted the situation, and many agreed that the road was necessary.[231]

To clear the approximately 90 houses and other structures from the right-of-way, the OOCEA entered into three contracts for demolition or removal and such related services as plugging water wells. Two of the contracts went to Chapman and Son, Inc., for $54,886.30 and $48,048, and the other was awarded to Freeman Truck & Equipment Company for $68,900.[232]

ZHA was named as general construction manager for the Western Extension in March 1988. Concrete girders and precast wall panels were purchased together with supplies for the Southeastern Beltway. In April 1988, the consultant selection committee of the OOCEA opened 15 proposals for CEI services on the two roads. Kaiser Engineering was selected to provide that

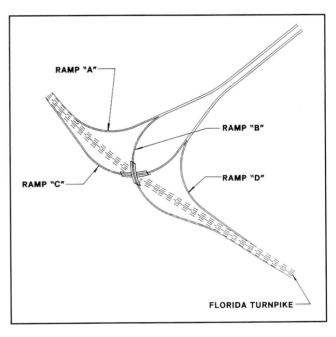

Design of East-West Expressway/Turnpike interchange. Only the northern section was built.

Looking west at construction of the Western Extension. At the top center Old Winter Garden Road veers to the right while the main line continues to the left.

service on Projects 50l and 502 for an amount not to exceed $1,900,000. Reynolds, Smith and Hills got the contract for Projects 504 and 505 for $1,700,000. CEI services on Project 506 (the toll plazas) went to Kunde, Sprecher and Yaskin for $600,000.[233]

With plans for construction to begin in June 1989, the first construction contract was awarded in late February to Asphalt Pavers, Inc., to build from west of Old Winter Garden Road to Good Homes Road for $6,364,388.81. The firm was given notice to proceed on June 1, 1989.[234]

Projects 504 and 505 (Good Homes Road east to the old East-West Expressway) were combined. Those projects were contracted in late February to Metric Constructors, Inc., for $18,641,179.10. A notice to proceed was forwarded on May 2, 1989.[235]

Project 50l was the interchange with the Florida Turnpike and the road from just west of Old Winter Garden Road. Metric Constructors was also the low bidder on that project and received the contract amounting to $5,442,231.65. Work began in late June 1989.[236]

Metric Constructors also got the contract to build the toll plazas for $2,793,700.[237]

Construction proceeded without interruption and the project was completed on October 8, 1990, two weeks ahead of schedule. With connections to the Florida Turnpike northward as well as to S.R. 50 via Good Homes Road, the Western Extension opened that same day without fanfare. OOCEA records show that the road cost $102,000,000. An estimated $32,000,000 of that was for right-of-way.[238]

Looking west at the intersection of the old and new roads in August 1989. Kirkman Road is at the bottom of the photo.

Looking east at the connection between the original East-West Expressway and the Western Extension at Kirkman Road in January 1990.

In 1991, a systems modification project (Project 510) was contracted to Martin Paving Company in the amount of $375,948.66 to modify the drainage, reconfigure the retention basin, and generally clean up the area west of Kirkman Road where the East-West Expressway had originally terminated at S.R. 50. At the other end of the project, according to the earlier agreement, the FDOT purchased a portion of the East-West Expressway at the Turnpike for $17,636,300.[239]

By the fall of 1990, the OOCEA had built four of the six roads which the 1983 Long Range Expressway Plan had recommended for completion by the year 2000. At a cost of about $385,000,000 it had added 26 miles of limited access toll roads and had done so on time and under budget. The East-West Expressway then ran nearly 25 miles from the Florida Turnpike to an intersection with S.R. 50 east of Alafaya Trail near Bithlo.

The Eastern Beltway was complete from the Bee Line north to S.R. 426 in Seminole County.

Two other roads, the Central Connector and the Southern Connector – an extension of the Eastern Beltway – were being planned, and discussions were underway about ways to finance and build the Western Beltway.

View eastward of the East-West Expressway with the completed Western Extension in December 1990.

Westerly view of construction underway at the intersection of the Western Extension and Hiawassee Road in December 1989.

View eastward from the completed interchange between the Western Extension and Florida's Turnpike.

Chapter 10

The Road Gets Rough:
The Central Connector

Planned as a 5.5 mile road connecting Interstate-4 near Michigan Street to the Bee Line west of Orlando International Airport, the Central Connector was intended to provide a limited access highway from downtown Orlando to the airport and reduce the heavy traffic volume on Orange Blossom Trail, Orange Avenue and S.R. 436. The Metropolitan Planning Organization had included the road in its Orlando Urban Area Year 2000 Needs Plan, and it was one of the six projects recommended for construction as a toll road in the OOCEA's 1983 Long Range Plan. In 1985, the MPO included the Central Connector in its Orlando Urban Area Year 2005 Financially Feasible Plan.[240]

In November 1987, the OOCEA board asked the staff to conduct preliminary feasibility studies of a Central Connector Project. In the meantime, the FDOT had disapproved an interchange with I-4, so it was decided to terminate the planned road at Michigan Street. A study by Vollmer and Associates concluded that the road was financially feasible even without the I-4 connection.

Deciding to go ahead with plans for the Central Connector, the OOCEA met in April 1988 to adopt it as its next project. More than 150 people crammed into the OOCEA's meeting room to protest the road. Fire Department personnel thinned the crowd because of fire regulations, but did nothing to lessen the intensity of the complaints from those who remained. Despite the protests, the road was approved for planning and construction.[241] Pursuant to that decision, PBS&J was instructed to complete a detailed financial feasibility study including engineering cost estimates.[242]

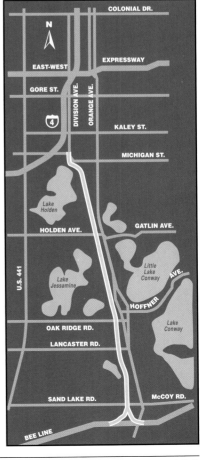

Central Connector as planned to connect the Bee Line with Michigan Avenue without an I-4 connection.

After the raucous April meeting, it was almost as if there were two parallel stories of the Central Connector. On the one hand, the OOCEA, long accustomed to opposition to its plans, continued with the necessary steps toward eventual construction, firm in the belief that the opposition could be satisfied or overcome. On the other, there was a determined opposition using whatever tactics were available – rhetorical, political, or legal – to resist the road. Some opponents just did not want it built. Others had clearly stated objections such as the lack of an I-4 connection, which seemed to them to make the road less useful. But, when that problem was resolved to the satisfaction of some opponents, others continued to resist as if defeating the road had become a cause in itself.

In August 1988, PBS&J delivered a detailed financial feasibility study including engineering cost estimates. The proposed road was expected to cost about $213 million, or about $40 million per mile, figures which the Orlando Sentinel reported time after time over the next several months. That same month the OOCEA approved a bond issue for the 1988 Project, the major part of which was the Central Connector. The plan was to sell the bonds in the fall, complete the necessary steps to begin construction in July 1991, and open the road in February 1993.[243]

In early September 1988, the South Orange Community Council, Inc., chaired by James Muszynski, described by the Orlando Business Journal as "a South Orange County activist," filed suit in the Orange Circuit Court to stop the bond sale.[244] Among the 19 allegations was a charge that "the strategy of the Expressway Authority is to keep the public in the dark as much as possible."[245] Judge Emerson Thompson dismissed the suit, and a sale of $140,600,000 worth of OOCEA bonds was completed in late October.[246] Muszynski declared that his organization would not give up just because it had failed to stop the bond sale.

Accustomed to dealing with such roadblocks, the OOCEA proceeded with its plans. In early October, John Dittmeier gave the OOCEA board a status report on the Central Connector Project. Among other things he said there would be an intense public awareness program, including three public information sessions, a public alternatives meeting, and a public hearing for right-of-way reservation. Documents would be available for viewing at the OOCEA offices, the consultant offices, and two public libraries. There would also be monthly newsletters for anyone wishing to be placed on the mailing list. At the same meeting, Brent Lacy of Barton-Aschman Associates, who was doing the PD&E study, added that his firm would also conduct at least 20 informal meetings in the neighborhoods in the area of the project.[247]

John Dittmeier, OOCEA director of planning, speaking to citizens interested in the Central Connector Project. (Photograph courtesy of the Orlando Sentinel.)

Those meetings were not likely to go unnoticed. That evening, following his report to the board, Dittmeier met with about 70 people who vigorously protested the road project. He told them that, "I know that it may be difficult to accept, but we are asking you to work with us on it."[248] In its report of the meeting, the Orlando Sentinel concluded that the Central Connector "will be open to traffic by February 1993 regardless of how many south Orlando residents rise up against it."[249] Already angry about the proposed road, few south Orlando readers distinguished between what Dittmeier said and what the newspaper concluded.

The corridor selected for the road ran south from Michigan Street, veering eastward around Lake Holden, then southward between Orange Avenue and the CSX Transportation railroad tracks, and ending at the Bee Line. There was some flexibility within the corridor, but all alternatives would go through Edgewood, a small incorporated city which stretched for about a mile along

Orange Avenue near the intersection of Holden Avenue. With a population of about 1,000, the city had a few hundred houses and many businesses along or near Orange Avenue, ranging in size from Wayne Densch's beer distributorship to small restaurants. Residents of the Lake Holden area near Edgewood and Belle Isle to the south also opposed the road. The South Orlando Business Group, Inc., joined the opposition, primarily because its members feared the lack of an I-4 connection would cause the road to inundate their businesses with traffic.250

It was Edgewood, however, which presented the most sustained, and eventually most effective, opposition to the road. When the Orlando-Orange County Expressway Authority was created in 1963, Winter Park had just recently emerged from a struggle to keep I-4 from being built through it. At the request of Winter Park officials, the local legislative delegation had managed to have included in the statute a provision which prevented the OOCEA from acquiring right-of-way within the limits of any municipality for any project without the approval of that municipality's governing body.251 When Edgewood's attorney discovered that provision, the small city became a large problem for the OOCEA.

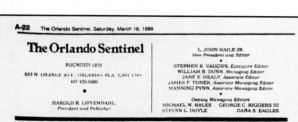

At a January 17, 1989, meeting, OOCEA Executive Director Bill Gwynn discussed the proposed road and the prohibitive legislation with about 120 concerned residents. According to the Orlando Sentinel, Gwynn, referring to the fact that the prohibitive statute did not apply to the FDOT, told those assembled that "You're going to have a road whether the city builds it or the state builds it whether you like it."252 Gwynn later said that he was misinterpreted – and the statement does seem to be incomplete – but the Sentinel version is what people remembered. Then, at the request of the OOCEA,

Representative Tom Drage tried unsuccessfully to have the prohibitive statute changed at the next legislative session.[253]

By April 1989, the David and Goliath analogy was being effectively encouraged by Edgewood Mayor Dorris Bobber and her allies, and public opinion was swinging behind Edgewood. Beyond that, other municipalities were reluctant to see local powers taken away. Over the objections of Orlando Mayor Bill Frederick, who strongly favored the Central Connector, the Tri-County League of Cities, made up of the mayors of eight Orange County cities, resolved to oppose the road and any changes to the law. Both the Orange County Homeowners Association and the Orange County Democratic Executive Committee followed suit. Conversely, resolutions favoring the Central Connector were passed by the Greater Orlando Chamber of Commerce, the Economic Development Commission, and the Downtown Development Board.[254]

Through the spring and summer of 1989, Gwynn and other OOCEA officials and Ed Skau of Barton-Aschman met repeatedly with the public to discuss the design of the Central Connector. Several changes were made in response to specific complaints.[255] But most of the opposition was directed toward any road, not just specific portions of it. In April, Mayor Bobber wrote a "My Word" article to the Sentinel. She recounted the many arguments, some of which were exaggerated or even erroneous, which had been made against the OOCEA and the Central Connector, but she emphasized two premises upon which she and many of her allies were acting. "The crux of the matter is that the city of Orlando and development interests want this road. The Expressway Authority has followed their lead, doing everything in its power to build the connector whether it's needed or not," she wrote. And, "...the authority has adopted a hard-sell approach, trying to intimidate Edgewood."[256]

Apparently agreeing with Bobber's last point, OOCEA Chairman B. J. West mailed a letter to 500 Edgewood residents just a few days before the agency adopted its preferred alignment for the Central Connector. West said in part, "...we have stepped on toes in the city of Edgewood. For that...I sincerely apologize."[257] A week later the OOCEA adopted an alignment which would have ended the road on its north end at the intersection of Division and Gore Streets. It would have taken 62 homes and about 130 businesses. None of the homes and only 19 of the businesses were in Edgewood.[258]

The next step for the OOCEA was to begin acquiring right-of-way, but it needed approval of the city of Edgewood to do that. When the matter came before the Edgewood city council on September 12, it was rejected by a 4-1 vote. Although the crowd of 150 cheered Mayor Bobber's speech, sentiment was by no means unanimous. Of the 21 people who spoke, 11 opposed the road, eight favored it, and two had no opinion.[259]

Unable to compromise with Edgewood, the OOCEA had two options, both of which would delay completion of the road and drive up its price. It could sue on grounds that the law was unconstitutional, or it could ask the state, which was not bound by the law, to build the road. In the meantime, plans for the road continued. A public hearing was conducted on the preferred alignment on October 10 at Oak Ridge High School. Contracts for final design consultants were awarded in November. Section I was assigned to DeLeuw Cather & Company for $2,250,000. Section II was contracted to Professional Engineering Consultants, Inc., for $1,470,000, and Bowyer-Singleton & Associates, Inc., agreed to design Section III for $1,210,000. A contract for Figg & Muller to design Section IV was approved but not completed as of November 22, 1989.[260] In April 1991, ZHA was selected to continue to provide construction management services for the Central Connector and the Southern Connector for a fee of $6,900,000. Eight engineering firms were chosen in June to perform CEI services on the two projects.[261]

In the meantime, the OOCEA sued Edgewood on grounds that the restrictive portion of Section 348 of the Florida Statutes was unconstitutional. That suit was dropped when the FDOT agreed to try to build the controversial road. Then Edgewood sued the OOCEA, asking that the building of the Central Connector be stopped. The South Orlando Business Group sided with Edgewood and spent some $80,000 for advertising against the Central Connector.262

While the suit was proceeding toward trial, the FDOT reversed its earlier decision and agreed to help obtain an I-4/Central Connector interchange. At that point, the South Orlando Business Group withdrew its opposition, saying that, "We believe that with this connection to I-4 the road works, . . ."263 The Orange County Commission was also supportive. But it had no effect on Edgewood. Mayor Bobber and other Edgewood residents had earlier opposed the road because it would not connect with I-4, but they continued their fight anyway.264

Hoping to have the I-4/Central Connector in operation by 1995, the agency initiated the process of designing it and preparing an Interchange Justification Report to obtain the necessary Federal Highway Administration approval. The PD&E study was contracted to Greiner, Inc., in August 1990 for $1,500,000. Public information sessions were held in April and June 1991, and the study was completed in August. DeLeuw Cather & Company was selected for final design of the interchange for $7,000,000 while the portion of the road from Michigan Street to the interchange was to be designed by Figg Engineers, Inc.,

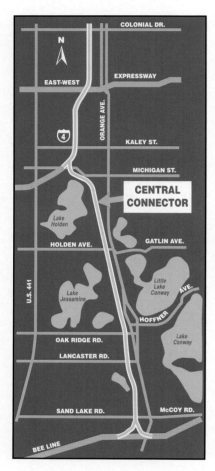

Map of Central Connector running between Bee Line and I-4.

at a contract price of $1,325,000. The Interchange Justification Report was completed and forwarded to the FDOT, but that agency never transmitted it to the Federal Highway Administration.[265]

Plans were underway at the same time to have the FDOT build the road, but bringing those plans to fruition was complicated if only because legislative approval was required. Most members of the local legislative delegation were more interested in killing or at least delaying the road than in seeing it built, but they were outmaneuvered by the last minute lobbying efforts of Orlando Mayor Bill Frederick in the 1991 legislative session. The FDOT plan to build the Central Connector survived in 1991 but it had far to go before construction could begin.[266]

Edgewood's suit against the OOCEA was tried before Orange Circuit Judge Frederick Pfeiffer in September 1990. Pfeiffer ruled that the city could not veto a road of regional importance (such as the Central Connector) despite the law giving local government the right to approve such projects. An appeal was filed with the 5th District Court of Appeals almost immediately. Nearly a year later, in August 1991, the appellate court overturned Pfeiffer's decision, ruling that Edgewood did have the power to stop the road within its city limits.[267] The adverse court decision was the turning point.

The OOCEA was still planning the Central Connector and the I-4 Interchange when the 5th Circuit Court declined to reconsider the case in October. It had also been continuing to try to meet with Edgewood officials to work out an agreement. But after the appellate court ruling, the OOCEA board discussed the possibility of abandoning the project.

At the November 13, 1991, board meeting, the members discussed "how it is determined if the Authority has exhausted all reasonable means to complete the Central Connector Project."

Linda Chapin requested that an expression of intent be made that day on the issue of the Central Connector, then followed several interesting comments from citizens in attendance.

Sandy Smith, whose business was in the path of the road, said he had been in a "holding pattern for more than four years" and he hoped to see something done.

Pat McCarey, who lived on East Gore Street, suggested that the OOCEA continue to try to work with Edgewood. Bob Harrell responded that the OOCEA had made many attempts to negotiate with the city, but it was opposed to the Central Connector under any conditions. Craig Andrews, an Edgewood Councilman said there was no change in their position.

Then Cecil Moore, an Edgewood resident, stated that "there are many people in the city of Edgewood that support the Central Connector," to which Mayor Dorris Bobber made a remarkable reply.

The mayor told those assembled, including her constituent, that "the city of Edgewood's stance on the Central Connector does not necessarily reflect its citizens' opinions on the Central Connector. Edgewood had the legal right to stop the Connector and did so on behalf of the citizens in the area that do not want it, not just those in Edgewood." After that discourse on virtual representation, the OOCEA apparently decided that it had "exhausted all reasonable means to complete the Central Connector Project." It passed a resolution that "We find it not possible to complete the Central Connector and urge legislative approval of the Western Beltway so as to allow it to become the next Authority project."[268]

At the next meeting, Acting Executive Director Joe Berenis reported that the three contracts for the road south of Michigan Street were complete and that Project 704 (north of Michigan) and Project 705 (the I-4 Interchange) were less than 60 percent complete. It was decided not to terminate those two contracts until it was known whether the FDOT would build the road, but to inform the contractors that their work might be terminated at 60 percent completion.[269]

Of the $140,600,000 borrowed for the 1988 Project, $75 million was paid off in July 1992, and the remaining $65 million was defeased in April 1993. When all contracts were completed or terminated, $20,048,125 had been spent on the Central Connector Project.[270]

When it resolved to abandon the Central Connector Project, the OOCEA asked the FDOT to take it over. The FDOT was willing, but it could do nothing unless the legislature approved. That turned out to be an insurmountable obstacle because Polk County Representative Fred Jones, who chaired the House Transportation Committee, wanted the FDOT's Turnpike District first to build an expensive limited access road around Lakeland in his home county. That road was placed on the legislature's list of roads, leaving insufficient funds for the Central Connector Project. Even metropolitan Orlando's local legislative delegation was little help. Some members were ambivalent, but others were determined to stop the road for a variety of reasons. The road was debated in the 1992 and 1993 legislative sessions, but nothing happened.[271]

In 1994 the Metropolitan Planning Organization formed a committee to review the transportation challenges facing the area between downtown Orlando and the Bee Line Expressway. The mediation process resulted in a final report entitled Surface Transportation Alternatives for the Orlando Urban Area South-Central Corridor Mediation out of which the Central Connector project was deleted from the region's transportation plan. Expressway Authority Executive Director Harold Worrall voted to support the decision even though it was a political decision and one based on a technical analysis of the needs." I agree to vote with the majority to eliminate the Central Connector not because I believe it is not needed, but simply because it is not going to happen and that it is time that the great dissension within our community end."[271A] Edgewood had won the fight to keep the road out of its city, but it could not keep the cars off Orange Avenue which continued to be jammed daily. Meanwhile the OOCEA was encountering other challenges.

Chapter 11

Building Roads and Mending Fences in the Early 1990s

Having just added 26 miles of new toll roads to the expressway system by 1990, the OOCEA was looking forward to its next road projects. But the unexpected and unrelenting opposition to the Central Connector was a serious setback. At the same time, the agency was obliged to deal with sensational reports of the peripatetic work habits of its chairman, B. J. West; the adverse reaction to the July 1, 1990, toll increase; a national economic downturn from 1990 to 1992; and public confusion over directional signs on the expanded expressway system. Although varying considerably in scope, each of these problems was addressed and resolved. In the long run, both the expressway system and the public perception of the OOCEA's methods of operations were improved, but in the meantime they made the early 1990s a difficult period for the agency.

Chairman B. J. West, the OOCEA's Operational Methods, Public Criticism, and Procedural Reform

In 1965, the OOCEA was managed by Richard L. Fletcher from a donated desk located in his insurance office. The agency moved into its own offices at 525 South Magnolia in the early 1970s, but James Greene was still doing most of the work with a two-member staff. In the mid-1980s much of the work formerly done by Fletcher and Greene was delegated to Executive Director Bill Gwynn and his small staff. The OOCEA had only eleven employees by 1990.

As the scope of its work grew, the agency adopted the requisite methods for building $385 million worth of roads ahead of schedule and under budget, but it had not kept abreast of changing public attitudes regarding the conduct of public agencies and their members. The Sunshine Amendment, ratified in the 1970s, required that meetings be open to the public. Equal Opportunity legislation required demonstrated efforts to make jobs available to minorities and women. Ethics codes, by which the conduct of public officials could be judged, were becoming commonplace. The effect of all these changes was that public agencies were required to prove that they were in compliance. That in turn necessitated written policy statements, procedural manuals, job descriptions, codes of conduct, and open meetings when decisions were being made.

The OOCEA had developed procedures which worked quite well in getting roads built, but it had concentrated entirely on results rather than procedures. Given the intense scrutiny to which the OOCEA was being subjected by such groups as James Muszynski's South Orange Community Council, and the Orlando Sentinel's reports of Chairman B. J. West's activities, those procedural shortcomings were soon to become major public issues. B. J. West had been, and in 1990 still was, a successful fund-raiser for incumbent Governor Bob Martinez. In 1987, he had appointed her to a special paid position in the Florida Department of Health and Rehabilitative Services to raise funds for its children's programs. He had also appointed her to the OOCEA board, an unpaid position. She was elected chairman of that agency in 1989. And Governor Martinez was running for re-election. While working for the HRS out of her son's real estate office and raising money from volunteers for children's programs, she was also the chairman of the OOCEA which selected contractors for construction work worth millions of dollars, and she traveled extensively on agency business. And there was the governor's re-election campaign. West told a reporter that "I'm able to squeeze everything into my schedule. I get up early and I work late. And I always have."[272] With her avowed work ethic, her preference for results rather than process, and wearing many hats, she was a likely target for investigative reporting.

After what it described as a two-week investigation, the Sentinel reported in

January 1990 that West had intermingled her HRS job, her OOCEA chairmanship, and her political activities. The result was that telephone calls for one activity were billed to another, work for one agency was performed while records showed her busy at the other, and there was considerable billing for overtime and at least one case of double-billing. Having never considered reporting in at eight o'clock and signing out at five as a day of work, West willingly acknowledged that "When I take [an expressway trip] that also gives me contacts into the private sector, it may very well be that it has advantages to HRS."[273]

Some would call that "networking," but when the gubernatorial fund-raising was added, others thought West had gone too far. Muszynski's South Orange Community Council, now being described by the Sentinel as "a transportation advocacy group," declared her activities "obscene," and called for her immediate resignation.[274]

Ignoring such critics, West instead called for an audit of her records by KPMG Peat Marwick, the agency's independent auditors. But that only brought more criticism because the KPMG Peat Marwick, partner-in-charge of the OOCEA account, had contributed to the Martinez campaign by buying tickets to an event which West had organized. Referring to that connection, Common Cause suggested that the audit be reassigned.[275]

Other allegations arising from that same fund-raiser had more serious implications for the OOCEA. The Sentinel reported that West had steered large contracts to at least 22 engineering firms, whose officials had contributed $57,000 to the Martinez campaign. Representatives of 12 of them attended the fund-raiser and sat at West's assigned tables. The charge, later challenged by Executive Director Bill Gwynn, was that West sat with Gwynn, Joe Berenis, and Greg Dailer on a committee which selected contractors at meetings which were closed to the public.[276]

Many more stories appeared in the Sentinel and some were carried in newspapers from Miami to Tallahassee, but the issue was whether West and several staff members had solicited funds from firms doing business with the OOCEA and led them to believe that successful bids were dependent upon whether contributions were made. When the newspaper sought information from officials of the firms, some declined to comment on grounds that what they said might affect their future business, but others said they had never felt any pressure or noticed any correlation between contributions and successful contract bidding. Several commented that it was just a way of doing business and they contributed in states where they did business as a matter of course.[277] But there was one exception. Kimley-Horn and Associates, Inc., a Raleigh, North Carolina, firm, reported to the Florida Department of Law Enforcement (FDLE) that an official in its Orlando office had declined a solicitation for a contribution and had been told that the action would affect the firm's ability to obtain OOCEA contracts. The FDLE then launched its own investigation.[278]

Based on information from the FDLE as well as the state comptroller, Orange-Osceola State Attorney Lawson Lamar took the matter to the Orange County grand jury in October 1990. After hearing witnesses from both HRS and the OOCEA, the jury indicted West on charges of grand theft and misuse of office, because she had conducted private and political business while she was supposed to be working for the Department of Health and Rehabilitative Services. It found no evidence that OOCEA contracts had been awarded on the basis of contributions to the Martinez campaign. It did say, however, that the practice of soliciting campaign contributions for Martinez from firms doing business with the OOCEA "gives rise to an inference of corruption."[279] It strongly urged reforms in the management practices of both the OOCEA and HRS. And it further asserted that the OOCEA was violating the Sunshine Amendment by holding closed meetings to select contractors.[280]

Three days after her indictment, West resigned from the OOCEA, but kept her HRS job until January 1992. In October 1991, she entered a plea of no contest to the charges against her and was sentenced to 100 hours of community service. In December 1990, the FDLE dropped its investigation of the Kimley-Horn allegations against the OOCEA, saying that it could find no corroboration of the charges.[281]

In December 1990, just before he left office, Martinez appointed Robert Mandell, another supporter of his re-election campaign, to succeed West. Thomas Barry had succeeded Ben Watts as the FDOT representative on the board, and Linda Chapin became the Orange County member in early 1991. Bob Harrell returned to the chairmanship, and Wilbur Gary remained treasurer. Mandell was made vice chairman.[282]

One of the first acts of the newly constituted board was to question the OOCEA staff's selection of a contractor for work on the I-4/Central Connector Interchange. One of the 19 bidders was Kimley-Horn. Feeling that their recent relationship with that firm would make their decision suspect, Bill Gwynn and his colleagues had decided to review the other 18 firms themselves and have PBS&J perform the evaluation of Kimley-Horn. Newly appointed member, Robert Mandell, suggested that the firm had not been treated fairly and called for an independent investigation. Linda Chapin, the other new member, agreed, saying "this is a disaster."[283] After a lengthy discussion, with Harrell and Gary dissenting, Thomas Barry joined the two new members in a 3-2 vote for an outside review. Composed of Robert Haven, Orlando's chief administrative officer, Orange County Public Works Director George Cole, and Thomas Barry of the FDOT, the committee concluded that Kimley-Horn had not only been treated fairly, but that the OOCEA staff had gone even farther than necessary to be fair.[284]

Building A Community

Whatever its purpose, the Kimley-Horn investigation was only an ad hoc occurrence. Over the next several months, on its own initiative, and later in response to a critical state Auditor General's report, the OOCEA adopted an extensive reform of its procedures. The net result would be that the OOCEA, with a demonstrated ability to build and operate its expressway system, would be in full compliance with recent legislative and constitutional changes respecting public agencies and would have the papers to prove it.

In February 1991, a five-member committee was appointed to review the OOCEA's policies and procedures and make recommendations for reforming them. Chaired by Russell Mills, president of Dyer, Riddle, Mills & Precourt, its other members were John Lowndes, president of Lowndes, Drosdick, Doster, Kantor and Reed; Robert D. Martin, chairman of Martin Paving Co., Thomas Lewis, vice president of Disney Development Co., and Kevin Walsh, coordinator of the Minority Business Enterprise office in Orlando.[285] The committee interviewed the OOCEA staff and PBS&J personnel, gathered information from other government agencies about their consultant selection processes, and sent questionnaires to firms which had done business with the agency. It then spent nearly seven months reviewing the information and developing a lengthy set of recommendations.[286]

The committee received 21 responses from 26 engineering firms which had previously done business with the OOCEA. Twenty of the 21 firms said the OOCEA selection process met or exceeded their expectations. The only dissenter was Kimley-Horn and Associates. The <u>Sentinel</u> concluded that the "Authority's method of awarding engineering contracts apparently is not as big a concern among engineers as it was with the grand jury."[287]

It was not the same, however, with smaller firms who had tried unsuccessfully to do business with the OOCEA. Nine of 14 firms owned by minorities and women expressed dissatisfaction with the way the agency did business. In its defense, OOCEA staff noted that minorities and women had received 13 percent of $19.4 million worth of projects during the past six years. But Chairman Bob Harrell said the agency could and would do better.[288]

In June 1991, months before the review committee reported, the OOCEA established a Disadvantaged Business Enterprise (DBE) policy and set a goal of awarding at least 15 percent of construction work to women and minorities.[289]

Submitted in September, the review committee's report included several specific items, but its general recommendation was that the OOCEA "develop and adopt written guidelines and procedures for all areas of its activities related to the selection and use of design engineers, appraisers, construction management engineers, MBE/WBE/DBE (Minority, Women's, and Disadvantaged Business Enterprise), contractors or suppliers, attorneys, financial consultants, contract employees, general design engineers and general construction engineers. Although the OOCEA had already opened all its meetings to the public, the committee further recommended the dissemination of information about how firms could go about doing business with it.[290]

After some discussion and some minor modifications, the operational review committees recommendations were accepted and implemented. Soon there were right-of-way procedures manuals, personnel manuals, and others spelling out the processes by which the agency operated. For example, the Consultant Recommendation Committee Evaluation Form Criteria detailed the criteria to be used for consultant selection. The OOCEA board also changed the composition of the selection committees to include the executive director and two other OOCEA representatives, and one representative each from Orange County and the city of Orlando. The committee would recommend the three top firms for board approval.[291] With several improvements such as that, the overall result of the adoption of the reform review committee was the formalization of OOCEA procedures in all phases of its activity.

The previously adopted goal of awarding 15 percent of construction contracts to DBE firms was expanded to include architects, engineers, and other professionals.[292]

A recommendation that continuing consulting contracts be limited to specific terms affected PBS&J, the general engineering consultant, ZHA, the general project management consultant, and J. Fenimore Cooper, who had been OOCEA legal counsel almost continually since 1965.

With respect to PBS&J and ZHA, the OOCEA board voted to retain both until the Southern Connector was completed. Then in 1995, Requests for Proposals (RFP) for both positions would be advertised. New contracts would then be for four years, with two, two-year extensions if the OOCEA so chose.[293]

There was extensive discussion about whether legal counsel should be furnished in-house or by consultants as in the case of Cooper. It was ultimately decided to allow time for Cooper to "wind-down on-going activities" and then, after a new executive director was on the job, to advertise for new legal counsel.[294]

Bill Gwynn had resigned as executive director in May 1991, and Joe Berenis had been acting in that position since.[295]

While the review committee was still gathering information, the OOCEA received an audit from the state's Auditor General, citing a lack of internal controls relating to selecting consultants, acquiring property, and maintaining records of expenditures. Some of the items noted had already been corrected, and others related to matters were being addressed by the review committee, but the audit required a response. After receiving the audit in July 1991, the staff worked individually and in committee through March 1992 responding to each of its specific recommendations.[296]

The state attorney's 1990 criticism of the agency's lack of a code of ethics had spurred action on that matter as well. After nearly three years, during which staff members were asked to familiarize themselves with publications about the Sunshine Amendment and Codes of Ethics for public employees, many meetings were held, advice was solicited from other agencies, and a draft was refined, and an Ethics Policy for the OOCEA and its employees was adopted in 1993.[297]

Toll Increases, Public Reaction, and Restructuring

With tolls scheduled to rise on July 1, 1990, to 75 cents at barrier plazas and 50 cents at exit ramps, the OOCEA contracted Orlando public relations firm, Fry, Hammond & Barr, to assist it in informing the public of the upcoming changes as well as its future programs. By that time, the South Orlando Sun was already blaming the increase on the proposed Central Connector. Then Joe Durek, an Orange County Commission candidate, sued the agency for spending money on the public relations campaign.[298]

Beginning on July 1, 1990, a more urgent form of opposition began when thousands of drivers avoided the expressway. There were also numerous verbal protests. On September 25, some 20 local trucking firms emphasized their displeasure by boycotting the expressway and crowding on to S.R. 50. The cities of Winter Garden and Ocoee passed resolutions asking the OOCEA to reduce the tolls.[299]

The increases had made tolls on the OOCEA system among the highest in the nation. Even the agency itself acknowledged that using the toll roads only paid off for those earning $10 per hour or more. "It kind of gives you the impression the expressway isn't for the poor," said Wilbur Gary, a nine-year member of the OOCEA who lived in west Orlando. He added that, "We may have to address it further."[300]

Concerned about the high rates as well as the loss of ridership, the OOCEA directed Vollmer and Associates to study what the impact of lowering the tolls might be on the expressway system. The firm undertook a study of the opinions of both users and non-users. It also attempted to identify places where toll adjustments might be implemented. PBS&J was subsequently asked to supplement that study with an origin and destination

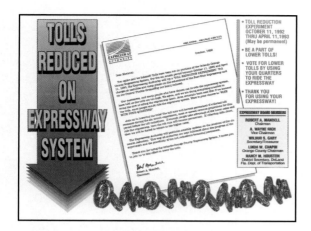

Flyer explaining the toll rate reduction experiment.

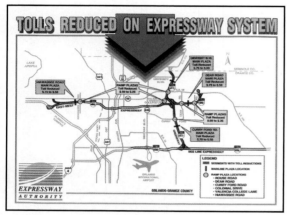

Map on back of flyer showing the plazas and ramps where tolls were being reduced.

Signs such as this kept drivers informed about the results of the experiment and encouraged continued use of the toll roads.

Motorists were informed that the toll reduction experiment was successful.

survey.[301] Dan Greenbaum of Vollmer and Associates reported in July that about 65 percent of non-toll road users polled did not ride the toll roads because of the high rates. Beyond that, a great majority of those polled on both the toll roads and the toll-free roads felt that the rates were unfair.[302]

An experimental toll rate reduction was the first priority of new Executive Director Harold W. Worrall, who joined the Expressway Authority in May 1992. Worrall had been recruited from the Florida Department of Transportation in Tallahassee where he had been serving as Assistant Secretary of Finance and Administration. By September of that year, Worrall convinced the Expressway Authority Board to try reducing toll rates at four toll plazas for six months. Tolls would be reduced from 75 cents to 50 cents at the University, Dean, Curry Ford, and Hiawassee Main toll plazas. And several exit ramps would be reduced from 50 cents to 25 cents. Meanwhile tolls would remain at 75 cents at the Holland East and Holland West main toll plazas on the East-West Expressway in downtown Orlando. It was hoped that enough drivers would use both downtown 75 cent plazas and to offset any revenue losses at the outlying 50 cent plazas. If the experiment was deemed successful, the reductions would be made permanent. To keep the bond holders from becoming concerned, the OOCEA set aside a reserve fund of $1,500,000 to be used in the event of a shortfall. Jorge Figueredo, newly appointed director of communication and marketing, was charged with advertising and monitoring the trial.[303]

From September 1992 through February 1993, toll receipts from the trial were running at about 92.5 percent of the target. The experiment was extended briefly, and in May the receipts were slightly above target, making the average for the entire test about 95 percent. That was apparently close enough. At its June 17, 1993, meeting the OOCEA board voted to make the toll reductions permanent effective on June 23. That was the date when bonds would be sold for a general refinancing of the OOCEA debt.[304]

National Recession, Declining Traffic, and Debt Refunding

With its projected revenues rising to more than $50 million annually by the early 1990s, a favorable national bond market, and the need for funds for its next project, the OOCEA initiated plans for a bond issue with which to build the Southern Connector. At that time, its debt was about $600 million.

In June 1990, the agency requested the Division of Bond Finance to approve bonds in an amount not to exceed $1.5 billion. The amount was not for a specific project but "for whatever project(s) the Authority determines in order to improve and extend the System." The Leon County Circuit Court validated them in August 1990. Bonds for the 1990 Project (the Southern Connector) in the amount of $385 million were then sold, thus increasing the OOCEA's total debt to nearly $1 billion.[305]

The increase in tolls in July 1990 would have been more than sufficient to make up for the anticipated decline in traffic volume. But the recession which lasted from about 1990 through 1992 added measurably to the traffic decline and seriously affected total revenues. Revenues for 1992, for example, dropped from an anticipated $57 million to about $51 million.[306]

Although revenues were still sufficient to manage the debt and meet other obligations, the decline was enough to cause concern. When the Central Connector Project was abandoned in late 1991, the debt was reduced by defeasing about $31.5 million left over from the 1986 Project and $75 million from the 1988 Project.[307]

Then on June 6, 1993, $564,415,000 worth of refunding bonds were sold. The interest rate was 5.78 percent. On November 1, another $202,715,000 of refunding bonds were sold. The interest rate was about 5.6 percent. The two sales resulted in a net present value savings of $43,105,506.[308] The actual cash flow savings from Fiscal Year 1993 through Fiscal Year 2004 was $100,719,810.[309]

As of June 30, 1993, the OOCEA's debt was $887,690,000. The national economy was recovering rapidly, the toll restructuring was in place and revenues were rising.

The Central Florida GreeneWay, Toll 417 and Toll 408

When there was only the Martin Andersen Bee Line and the Spessard Holland East-West Expressway, it was a simple matter to follow the directional signs. But with the Western Extension and Eastern Extension of the East-West Expressway and an Eastern Beltway with northern and southern sections, signage became much more difficult. Added as the new roads were built, signs on the expanded expressway system were confusing. Members of the OOCEA board and its staff admitted to having been lost on the new roads.

Robert Mandell recommended that the Eastern Beltway, from I-4 in Seminole County to I-4 in Osceola County, be renamed the GreeneWay in honor of the recently deceased James B. Greene who had chaired the OOCEA from 1971 to 1985. The OOCEA board approved the suggestion. Naming a highway after an individual requires legislative approval, and of course the change would have to be approved by the several counties.[310] Those steps were taken in due course.

The renaming of the road in memory of James Greene was well-received, but it would not solve the signage problem. Bob Paulsen of PBS&J noted that the problem was that both the East-West and the Eastern Beltway had directions in their names. His firm proposed using a numbering system, adding that the East-West was already designated by the FDOT as S.R. 408 and the Eastern Beltway was S.R. 417. The Turnpike District, since it was funding the Beltway in Seminole County and the

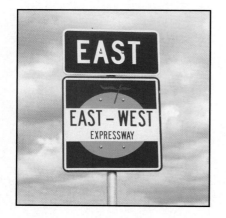

Signs such as this confused tourists, local drivers, and OOCEA board members alike.

The change to new signs using road numbers and destinations was expensive, but it made driving the Expressway system much easier.

Southern Connector Extension from S.R. 535 to I-4, had already agreed to use the numbers, and the FDOT had recently approved a new toll road symbol.[311]

Most OOCEA members preferred names over numbers, while the FDOT preferred numbers. Some wanted the Eastern Beltway to be named the Central Florida Greenway, but environmental activists vigorously opposed that name, arguing that it implied a natural pathway rather than a paved road.[312]

It was ultimately decided to name the Eastern Beltway, the Central Florida GreeneWay, but also to identify it as Toll 417 and the East-West Expressway as Toll 408. The new signs would use destinations rather than directions to direct traffic.[313]

The new signs were installed on the existing Central Florida GreeneWay by January 1993 and were incorporated in plans for the Southern Connector which was then nearing completion. The only parts of the system which were still to be converted at that time were the East-West Expressway and the Bee Line, the cost of which was estimated at $3.6 million for the East-West and $2.2 million for the Bee Line.[314]

Wilbur H. Gary, OOCEA board member, 1983-1993.

Personnel Changes: 1990-1993

When Robert Harrell resigned from the board in late 1991, Robert Mandell became chairman. A. Wayne Rich was appointed to fill the vacancy created by Harrell's departure and succeeded Mandell as chairman in 1992. In August 1992, Nancy Houston replaced Thomas Barry as the FDOT representative.[315] Wilbur Gary was succeeded by Inez J. Long in early 1993. At that time the board membership included Wayne Rich, chairman; Robert Mandell, vice chairman; Inez Long, secretary/treasurer; Linda Chapin, Orange County chairman, ex officio; and Nancy Houston, FDOT, ex officio.

Harold Worrall had become executive director in 1992 and Jorge Figueredo was shortly named director of communications and marketing.[316]

The new general counsel was Akerman, Senterfitt & Eidson with Tom Ross as partner-in-charge. The firms of Broad & Cassel and Perry & Lamb were designated to handle condemnation cases.[317]

Thomas Barry, FDOT, ex officio member of the OOCEA, 1989-1992. Barry became Secretary of the FDOT in 1997.

OOCEA Board of Directors in 1993. From left, they are Linda Chapin, Orange County chairman, ex officio; Inez Long, A. Wayne Rich; Robert Mandell; and Nancy Houston (seated), FDOT, ex officio.

Dr. Harold Worrall, OOCEA executive director, was FDOT Assistant Secretary for Finance and Administration before being recruited by the OOCEA Board in 1992.

Jorge Figueredo, director of operations, communication and marketing since 1992.

Chapter 12

The Southern Connector, the Southern Connector Extension, and the Seminole County Expressway

As it was finally designed for construction in the late 1980s, the Southern Connector was a 22-mile road running south and west around the airport from the Bee Line to S.R. 535/536. The longest toll road built in Orange County since the Bee Line in the 1960s, it was an extension of the Eastern Beltway – 15 miles of which was already built – which would ultimately extend nearly 60 miles from I-4 near Sanford to rejoin I-4 in the Disney World area. While it was still being designed, changes were made so that it would merge into a Southern Connector Extension which would run westward for several miles from S.R. 535 and intersect with I-4 south of Disney World. About the same time plans were being made for a 12-mile extension on the north end of the existing beltway in Seminole County. It would begin at S.R. 426, cross Lake Jesup and end at U.S. 17-92 south of Sanford. When these three projects were finished, the Eastern Beltway – by that time known as the Central Florida GreeneWay – would be complete except for a six-mile gap between U.S. 17-92 and I-4.

The idea of a "beltline project in the sector from the Bee Line to Interstate 4" was introduced in 1974 by Nelson Boice, then president of Florida Ranch Lands. He wrote James Greene offering to seek donations of right-of-way from landowners who would benefit from such a road through undeveloped southern Orange County.[318] At that time, the OOCEA, as agent for the FDOT, was studying the feasibility of a 33.5 mile road around the east side of metropolitan Orlando. After traffic and earnings studies showed in 1975 that such a road was not financially feasible, the offer of donated land languished. It was revived in the mid-1980s when four large landowners calling themselves the Southern

Connector Group, on their own initiative, contracted with Miller, Miller, Sellen, Einhouse, Inc., to do a corridor study of a 13-mile road that would link the airport with South Orange Blossom Trail.[319]

According to Jim Sellen of that firm, the plan was actually to build a road, leaving a center strip wide enough for an expressway to be donated when the OOCEA decided to build. Their road would then become frontage roads alongside the expressway. The bold offer was well-received. OOCEA Chairman Phil Reece said, "It's got to be a good deal," and Orange County Commission Chairman Tom Dorman added, "if we can get something built at no cost to the community, it's certainly something we need to look at."[320] But plans for the private undertaking were quite imprecise, and James Greene had noted earlier that "the concept differs from the OOCEA Long Range Plan and should be revisited."[321]

The Southern Connector had not been included in the 1983 Long Range Expressway Plan, but since 1981 Orange County had been busily approving large developments in the area such as Meadow Woods, Southchase, Lake Nona and Hunters Creek. By making its approval of most of those projects contingent upon adequate transportation, the county was increasing pressure for the Southern Connector.[322] By 1983, the GOAA was also urging extension of the Eastern Beltway to provide a southern access to the airport.[323] As late as September 1986, however, the Southern Connector was still "a long range concept at the Expressway Authority," according to the Orlando Sentinel.[324]

Then, at the urging of the Southern Connector Group in March 1987, the OOCEA added the road to its long range plan and agreed to include the group's $250,000 corridor study in its planning. At that time, the group was still agreeing to donate right-of-way, but the scope of the overall study had been expanded to include the entire route from the Bee Line to I-4.[325] In October, the Orange County Commission advanced $400,000 for a six-month study to designate a preliminary route for the road.[326]

At that point, the Osceola County Commission announced its interest, apparently wishing to have the expressway routed along Dart Road just south

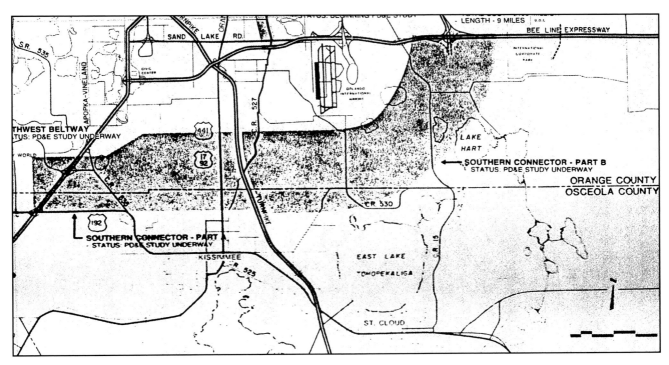

This corridor study for Parts A and B of the Southern Connector was completed in 1989. It recommended that the new road be located within the shaded area.

of the Osceola County-Orange County line. During the next seven months, OOCEA officials held many meetings with Osceola authorities who kept changing their position on routes and methods of financing. Negotiations continued through May 1988, but no agreement was reached.[327]

The first public meeting to discuss alternative alignments of both Parts A – from I-4 to Orange Avenue – and B – from Orange Avenue to the Bee Line – of the Southern Connector was held at the Tupperware auditorium on June 22, 1988. As of that date, busy with all four of the 1986 Project roads, the OOCEA did not have the financial capacity to build the Southern Connector.[328]

Revival of the Turnpike's Road Building Plans

In the 1960s, Charles Rex, chairman of the old Turnpike Authority, had begun implementing a plan not only to extend the Turnpike itself, but also to build other roads which connected with it. He was actually completing plans to build a connector from west of the Orlando International Airport to I-4 when the Turnpike Authority was abolished during the reorganizations following ratification of the 1968 Constitution. But in 1988, the FDOT obtained legislative approval to use Turnpike revenues to back bonds to finance a building program similar to the one Rex had previously initiated. Since these funds would have to be spent on roads designated by the legislature, the

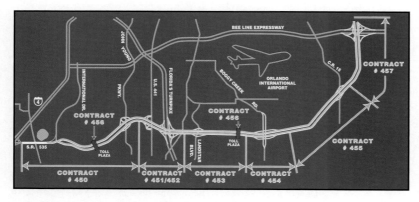

The Southern Connector as designed before the Turnpike District's extension was added.

renewed plan set off tremendous battles in that body, and between it and the governor, but it nevertheless offered the possibility of funding for metropolitan Orlando's Beltway which was otherwise unavailable.[329]

Orlando's hopes for an infusion of new funds to build its Beltway were thwarted in the legislature, but, in the meantime, the FDOT agreed to spend $37 million to build an interchange for the Southern Connector at its intersection with the Turnpike. Although the funds never materialized, their inclusion in the FDOT budget enabled the OOCEA to obtain financing for construction of the Southern Connector.[330] Beyond that, the 1990 legislature would authorize the use of Turnpike bond funds for construction of the 12-mile northern extension of the Beltway – also known as the Seminole One Project – from S.R. 426 to U.S. 17-92 in Seminole. Meanwhile, an innovative public-private venture would be undertaken to build the Southern Connector Extension through Osceola County.

Planning, Financing and Building the Southern Connector

While the OOCEA was holding public information meetings and preparing right-of-way maps for Part A of the Southern Connector, and negotiating with several large landowners along Part B, Disney officials proposed a change which transportation authorities thought made sense. As being planned, the Southern Connector would have converged with S.R. 535 and S.R. 536 as well as an extension of International Drive near the I-4 interchange at Epcot Center. The OOCEA, the FDOT, and Orange County transportation personnel agreed with Dave Grovdahl of the ECFRPC that adding the Beltway at that point would make it "a very busy place."[331] The Disney proposal was to extend the Southern Connector southwesterly into Osceola County for several miles to an interchange with I-4 west of Disney World. The extension made sense to transportation authorities because it would make the Southern Connector intersect with I-4 near the point of a planned intersection with the western part of the road, providing the access necessary for Disney to build the Celebration community. In October 1989, preliminary plans for the extension were announced. Greiner, Inc., was doing the design work. It was later revealed that Disney and three other large landowners were working quietly with Osceola County to finance the project.[332]

In the meantime, the OOCEA concentrated on a shortened version of Part A extending from about one mile east of S.R. 535 to Orange Avenue, and the original Part B from that point eastward to the Bee Line. Planning was more complex than usual because the OOCEA was working with design engineers on the Southern Connector Extension to the west, and Miller-Sellen Associates, the engineers employed by the Southern Connector Group, on the east. In addition, there were extensive negotiations with property owners other than the Southern Connector Group, the GOAA – because of the connecting road between the new expressway and the airport – and Orange County.[333] A centerline for Part A was accepted in August 1989, and one for Part B was approved in March 1990. Joe Berenis explained that, for both portions of the road, there would be interchanges at International Drive, John Young Parkway, U.S. 441, Florida Turnpike (which was the responsibility of the FDOT), Landstar Road, Boggy Creek Road, Narcoosee Road, and the Bee Line Expressway. Two mainline toll plazas would be located between International Drive and John Young Parkway and Landstar Road and Boggy Creek Road. Berenis added that the staff had been working closely with FDOT regarding the Florida's Turnpike Interchange and fully expected to open it concurrently with the Southern Connector.[334]

Contracts for preliminary engineering services were let to Greiner, Inc., for $660,000 and Dyer, Riddle, Mills & Precourt, Inc., for $750,000 in November 1989.[335]

View east across S.R. 535 and International Drive-S.R. 536 intersection at the beginning of construction of the OOCEA's portion of the Southern Connector.

Final design consultants were selected in December, and contracts were completed in March 1990. The road was divided into six sections beginning near S.R. 535 on the west end. Greiner, Inc., was awarded a contract for Section I in the amount of $2,450,000. Section II went to Howard, Needles, Tammen & Bergendorff for $1,625,000. Wilbur Smith and Associates agreed to design Section III for $1,125,000. Beiswenger Hoch & Associates was chosen for Section IV. The contract was for $2,125,000. Section V went to H. W. Lochner for $1,450,000, and Dyer, Riddle, Mills & Precourt was awarded a $1,850,000 contract to design of Section VI. In late 1990, toll plaza design was contracted to Metric Engineering, Inc., for the western section (Part A) and to Reynolds, Smith & Hills for the eastern section (Part B). The contracts were for $385,000 and $413,000 respectively.[336]

Wetlands Mitigation

The OOCEA had been working with the Department of Environmental Regulation to compensate for destroying wetlands by creating new ones and maintaining them for five years. Because the Southern Connector would involve destruction of 114 acres, OOCEA officials and environmental groups decided to seek a better way to satisfy environmental requirements on the one hand and preserve the environment on the other. Instead of creating wetlands alongside the new road, the OOCEA wanted to pay lump sums to the St. John's River Water Management District and the South Florida Water Management District – through which the Southern Connector would pass – with which those agencies could purchase large tracts of sensitive lands and save them from development.

Introduced by Representative Tom Drage of Maitland, the Beltway General Mitigation Bill was enacted in 1990 as Florida Statute 338.250. Although both the state and national governments still had to approve, the statute provided that, when the OOCEA received permits from the DER, it would then pay the water districts in lump sums. For the Southern Connector, the St. John's district was paid $11,280,000, and the South Florida district received $3,000,000. The districts would then use the funds to purchase such endangered wetlands as the BMK Ranch in Orange and Lake counties, Seminole Springs in Lake County, and land along the Econlockhatchee River in east Orange County. The only problem, soon to be revealed, was that the state, and not the counties, would select the properties to be purchased.[337]

The statute and the increasing complexity of environmental permitting requirements also increased other indirect costs. In 1992, Bill McKelvy told the board that permits issued by the DER and the Water Management Districts were becoming more and more complex due to the many special conditions required. Although the general engineering consultant and the CEI consultants provided daily monitoring, he requested approval to hire an independent firm which specialized in environmental sciences for all projects exceeding $1 million. It would be responsible for assuring the adequacy of pollution control features which were included in construction contracts. The board agreed, and in early 1993 BDA Environmental Consultants was chosen to perform the new service.[21]

Funding the Southern Connector (the 1990 Project)

To enhance the marketability of a $385 million bond issue for the 1990 Project, the OOCEA once again needed a pledge of gasoline tax funds from Orange County. Once again, the Orange County Commission raised objections to making that pledge. This time, the issue was the recently enacted wetlands mitigation bill which gave the state the final approval on which land would be purchased with mitigation funds. Because the land selected might not be in Orange County, Commissioner Vera Carter led the commission in demanding concessions from the OOCEA in return for the gas tax pledge. The list of demands was lengthy, but the intent was to force the OOCEA to assist in keeping the mitigation funds in Orange County. Unable to comply with such demands, the OOCEA rejected the conditions. Having approved development projects which would add perhaps 50,000 new inhabitants in the area of the planned road, the county commission had some obligation in the matter but

Building A Community

Looking north on Boggy Creek Road toward the Orlando International Airport before the bulldozers came. The Southern Connector right-of-way is in the center where the two clumps of trees are on the right.

held fast to its demands for three months. As Carter herself admitted, "We have approved subdivision after subdivision and the roads are already overloaded."[339]

The commission finally pledged the funds after the OOCEA agreed to give the county a voice in locating future beltway interchanges. This was not an onerous concession, since the agency had always coordinated such matters with the county.[340]

The bond sale for the 1990 Project was completed in December of that year.[341] An open house was held on February 7, 1991, to update the public on the Southern Connector. Final designs were expected to be completed in April, and right-of way acquisition would begin immediately. Construction would begin in late 1991, and completion was anticipated for July 1993.

Right-of-Way Acquisition

Right-of-way acquisition for the Southern Connector was a little different from the earlier projects, but no less arduous and certainly no less costly. Of the 142 parcels required, 112 of them were in the Project 450 area, the westernmost 4.3 miles of the road. They accounted for less than $5 million of the total right-of-way costs. There were only 30 parcels in the remaining 17.7 miles, but they cost about $60 million.

The difficulty encountered in Project 450 was an old subdivision, which had been abandoned after the collapse of the 1920s Florida boom. Locating the title holders of the numerous 50 foot lots in that subdivision was a challenge, but the ultimate costs were comparatively modest.[342]

Looking east across the John Young Main toll plaza under construction in April 1993. The Shingle Creek bridge is at top of picture.

The remaining 30 parcels were more difficult to obtain and far more costly. It turned out that all those offers of right-of-way from the Southern Connector Group and other large developers had not meant donations of land. Rather, they had meant that the land would be made available for purchase. Then there were disagreements over appraisals resulting in several costly law suits. The Lake Nona Corporation was awarded $8,048,290 for the necessary right-of-way through its land. In two separate suits, Southchase, Ltd. received $17,488,657. Landstar Development Corporation obtained a judgment for $1,350,000. James Forest (Sonny) Lawson received $7,431,173 as compensation for the portion of his land which was condemned for right-of-way. American Newland Associates, developers of Hunters Creek, waged an expensive battle in court before finally settling for $19,550,000. There were several smaller judgments. Kirby Smith Groves, Inc., for example, declined an offer of $210,678, went to court, and obtained a judgment for $613,617. Eschewing the courts, Charles E. Bradshaw, Jr. had long since worked out an agreement with Charles Sylvester and was paid $3,025,700 for his portion of the right-of-way.[343]

It was necessary to remove only one house along the entire 22 miles of right-of-way.

Although the sum does not include appraisers fees and some legal expenses, right-of-way acquisition for the Southern Connector amounted to about $65 million.

Looking east over the U.S. 441 (Orange Blossom Trail) intersection in April 1993.

View east over Florida's Turnpike and the railroad bridge in April 1993. Landstar Boulevard is in the upper left.

Building the Road

The existing contract with ZHA for general construction management was extended in February 1991 to include the Southern Connector. Three contracts for prepurchased materials were completed in June. Dura Stress, Inc., was paid $6,125,730 for prestressed concrete pilings. Precast retaining wall panels were purchased from VSL Corporation for $1,992,022. Trinity Industries, Inc., contracted to furnish steel bridge girders for $4,330,925.[344] Contracts for testing of materials were let to four laboratories for a total amount of $1,200,000.[345] Five firms were awarded contracts for CEI services at a cost of $12,782,000.[346]

Construction contracts were awarded in August and September, and the first earth was turned in early November 1991. Dayco Astaldi Construction Corporation, a South Florida firm, was chosen to build Project 450 for $28,027,298.56. The project started one mile east of S.R. 535 and extended eastward 4.3 miles. It included an interchange at John Young Parkway and a bridge across Shingle Creek.[347]

The contract for Project 451 was let to Martin K. Eby Construction for $17,083,848.87. That project included 2.5 miles of roadway with an interchange at U.S. 441.[348] Dayco Astaldi was also chosen to build Project 453 for a contract price of $13,061,535.26. It was an approximately four mile segment from Orange Avenue to just west of Boggy Creek Road and included a bridge over the CSX Transportation railroad tracks and an interchange at Landstar Road.[349]

Project 454 went to Hubbard Construction Company in the amount of $14,928,038.45. It was a 2.2 mile segment of road with an interchange at Boggy Creek Road, an overpass at the Orlando Utilities Company railroad tracks, and a bridge across Boggy Creek. It also included a

The Boggy Creek toll plaza is in the foreground and the Boggy Creek Road interchange is in the upper part of the picture. Photograph taken in February 1993.

portion of a new Boggy Creek Road which would meet another segment being built by the GOAA for a southern access to the Orlando International Airport.[350]

Project 455 was awarded to Bergeron Land Development, Inc., for $12,182,564.40. It was an approximately three mile portion of roadway including an interchange at Narcoosee Road.[351] Project 457, the last segment, was a 5.8 mile stretch running from near Narcoosee Road to the Bee Line. It included the half of the Bee Line Interchange which had been left incomplete when the Southeastern Beltway was built, an overpass at Moss Park Road, an overpass at the CSX tracks, and other bridges. The contract was let to Hubbard Construction for $23,916,919.08.[352]

Including those already mentioned, the entire 22-mile route required 50 bridges.

The six road construction firms received notices to proceed in November and December 1991.

Project 456 was let to Martin K. Eby for $8,525,233.20. It included two mainline barrier toll plazas and ten, two-lane ramp plazas. Notice to proceed was given in April 1992.[353]

All of the contractors were required to demonstrate the Disadvantaged Business Utilization for each of their projects.

In February 1992, with total construction work about ten percent complete, Bill McKelvy reported that DBE participation was about 16.57 percent, and DBE firms had been paid about $1,120,000 during the preceding four months.[354]

As director of construction, Bill McKelvy was usually in good spirits during the 19 months of the Southern Connector construction. Once the right-of-way problems were resolved, he had mostly open country through which to build. Beyond that, the weather was generally cooperative. In May 1993, he was almost guaranteeing that the July 1 completion date would be met.

The Southern Connector under construction between Boggy Creek Road and Narcoosee Road in October 1992. The Stanton Energy Plant can be seen in the upper left.

Some problems were encountered along the way, but only three of them were directly related to construction. With the need for 50 bridges and overpasses as well as large quantities of fill for the roadway across miles of low land, some 15 million cubic yards of dirt were required. It came from several sources and caused the usual complaints of noise, dust, and disruption of traffic, but the most serious problem developed in the area just south of the airport where Hubbard Construction was building Project 454.

The Narcoosee Road interchange in July 1993.

The difficulty arose over a six-acre borrow pit belonging to Maury Carter from which a million cubic yards of dirt was being taken. The problem was the heavy trucks speeding along Boggy Creek Road, State Road 15 (Narcoosee Road), and Cyril Drive. According to a Sentinel story "local residents continue to fear for their safety as the kamikaze dirt drivers hurtle down the county highways at speeds exceeding the posted limits." A number of the truck drivers were apprehended for speeding and Hubbard officials warned them that speeders would be fired, but they were private contractors ultimately responsible for their own conduct. The best solution was that Hubbard expected to complete the hauling project in early March, about a month after the story appeared.[355]

Another problem arose from Hurricane Andrew. Dayco-Astaldi, a South Florida firm, was using many trucks from that area on the two segments of the Southern Connector it was building. Many of the truckers left for Miami as soon as the hurricane passed over. Faced with a July 1, 1993, deadline for completing Projects 450 and 453, Dayco-Astaldi was obliged to pay premium rates to secure enough trucks to complete the jobs on time. It then asked the OOCEA for a supplemental sum of $2 million to compensate for the additional expenses. A settlement was reached in early 1994 by which the OOCEA paid the firm $1,150,000.[356]

Looking northeast across the Southern Connector/Bee Line interchange after construction was completed.

A more complicated problem resulted when Hubbard Construction Company submitted claims for additional compensation on Projects 454 and 457 amounting to nearly $5 million. The claims led to litigation which was settled only after lengthy negotiations during which the OOCEA seriously considered barring Hubbard from bidding on future contracts.[357]

The Missing Interchange and the Road that Arrived Late

The Florida's Turnpike Interchange with the Southern Connector had been authorized by the FDOT, and the OOCEA was acting as its agent in planning and designing it. It had even purchased some of the right-of-way. But, the FDOT decided that traffic projections were insufficient to justify the interchange at the time and withdrew from the project. The OOCEA completed the design work which was then in progress, but the Turnpike Interchange was left unbuilt. The lack of an interchange between two such important limited access roads was unfortunate.[358]

The GOAA had been anxious to have the Southern Connector constructed, because it could provide a much-needed southern access to the Orlando International Airport. But that required a new road to be built southward from the airport to connect with the Boggy Creek Interchange. Plans had been coordinated so that the OOCEA was providing a portion of the new road to which the GOAA had only to connect its part from the airport.

In the foreground is the Florida Turnpike without the interchange which had been planned.

The 2.7 mile road was considered essential to both the GOAA and the OOCEA. It was expected to relieve congestion on the Bee Line and S.R. 436 on the north side of the airport and increase the ridership on the new Southern

This view across the Boggy Creek interchange shows the OOCEA portion of the Boggy Creek Road ready for traffic. Construction is still underway on the OIA portion at top of picture.

Connector on the south. But the GOAA encountered difficulties in raising the $22 million necessary to build the road. Relief came in August 1992 with an $8.9 million grant from the FAA, enabling the GOAA to fill in the remainder from a state grant and a $3-per-flight user fee. But construction did not get underway until late 1992, too late to complete it by the time the Southern Connector opened.[359]

The Opening Ceremony and the 30th Anniversary of the OOCEA

With construction running nearly 10 percent ahead of schedule in May 1993, Bill McKelvy had assured everyone that, barring an unlikely rash of storms, the road would be completed as planned by July 1. He was correct. At a final cost of $273 million, the Southern Connector was completed on time and under budget. Because completion of the Southern Connector was coinciding with the Orlando-Orange County Expressway Authority's 30th birthday, a special celebration was planned for the opening of the road. Planned by Jorge Figueredo, the celebration was held on June 26. Secretary of State Jim Smith and OOCEA board members were on hand at the ribbon-cutting ceremony at noon, but the celebration had begun hours before that with a 5-kilometer walk,

The opening ceremony for the Southern Connector was combined with the celebration of the OOCEA's 30th birthday. It was sponsored by many of the firms which had participated in building the road.

The balloon rides were quite popular.

a 15-kilometer bicycle ride, a 10-kilometer skate, a moonwalk for children, and hot air balloon rides. There were free hot dogs for everyone. An antique car display featured cars of the 1963 vintage, emphasizing the 30th anniversary of the OOCEA.360

The Southern Connector tolls were one dollar at the two main barrier plazas and 50 cents at the ramps. Collections began on July 1, but results were disappointing. Executive Director Hal Worrall explained that it would take a few weeks for people to get acquainted with the new facility. It was also noted that delays in completing the airport access road were depriving the new road of about 25 percent of its anticipated traffic.361 The ridership was expected to increase in due course and, in the meantime, the OOCEA had the income to pay its bills.

At noon, Secretary of State Jim Smith spoke while A. Wayne Rich (left), Robert Mandell, Inez Long, and Nancy Houston listened.

Jim Smith joined A. Wayne Rich in cutting the ribbon.

The Southern Connector Extension

The Southern Connector Extension was finally designed as a 6.4 mile road beginning about a mile east of S.R. 535 and running southwesterly into Osceola County where it intersected with I-4 about four miles past Disney World. It was part of a larger public-private venture involving Disney World, three other large landowners, and Osceola County. These parties wanted to build a 12.4 mile Osceola Parkway from Florida's Turnpike near Kissimmee to Disney World, following the already existing Dart Road which ran just south of the Orange County-Osceola County line, as well as the Southern Connector.

The opening of the Southern Connector Extension on June 26, 1996, also attracted a crowd. The E-PASS booth, featuring the OOCEA's new electronic toll system, attracted considerable attention.

After nearly three years of negotiation, the parties signed a complex contract on July 24, 1992, by which revenue bonds would be issued to build the Osceola Parkway. The Reedy Creek Improvement District agreed to pay $1 million dollars per year toward managing that debt. Additionally, it agreed with the three other landowners to pay a specified annual sum if necessary to make up any shortfalls in toll revenues. If those sums were inadequate, then Osceola County would have to contribute a maximum of $1.3 million per year toward debt management.[362]

The Parkway was completed in August 1995, but it has not been the revenue producer that was anticipated.[363]

Part of the overall agreement, the Southern Connector Extension was handled in a different way. While Disney and the other landowners contributed significantly to the project, and Disney built a 2.5 mile frontage road along I-4,

Florida's Turnpike District built the main roads using Turnpike revenue bonds. All tolls collected on the completed road were paid to the Turnpike District.³⁶⁴

Right-of-way acquisition began in early 1994 and construction was underway by the end of that year. Construction in the busy S.R. 535 and S.R. 536 area was difficult, and there were some traffic delays on those roads, but progress was steady and the road was actually completed ahead of schedule. It opened on June 23, 1996. At a final cost of about $153 million, it was the largest public-private project built by the FDOT. The state agency spent about $79 million on the road, with the private parties paying the rest – nearly half of the total amount – in cash or in-kind contributions. An FDOT official said that "without that kind of cooperation, it wouldn't have been feasible for us to build it."³⁶⁵ It was a significant section of road because it connected the Central Florida GreeneWay with I-4 on the southwestern end. In the meantime, progress was being made toward a similar connection on the northern end.

The Seminole County Expressway

The Seminole County Expressway Authority had selected an alignment for the toll road through the county in 1987 and reserved right-of-way. It would begin at S.R. 426, pass between Winter Springs and Oviedo, cross the western end of Lake Jesup and C.R. 427 (Lake Mary Boulevard), cross U.S. 17-92 south of Airport Road, cross Vihlen Road and then curve westward to I-4. The route was about 17.7 miles long. When traffic studies showed that only the portion from S.R. 426 to U.S. 17-92 (about 12 miles) was financially feasible, the SCEA contracted with six engineering firms for final design of that part of the road. Encountering difficulty in financing for even that 12-mile stretch, the SCEA was pleased when the 1988 legislature authorized the use of Florida Turnpike revenue to back bonds for road building. Then, because of the 1989 stand-off between the Governor and the legislature, the SCEA decided to go it alone. A bond issue was prepared, complete with a pledge of county gasoline tax funds. The idea was to build the entire four-lane road if funds were sufficient. If not, part of the 12-mile road would be only two lanes. But, in

Anticipating completion of the Seminole One project soon after the Southern Connector opened, both the Florida Turnpike District and the Seminole County Expressway Authority had booths at the June 26, 1993, dedication ceremony for the Southern Connector.

1990 the legislature and the Governor agreed on a measure which authorized $176.7 million for the road. Obliged to choose between the certainty of state funding or take a chance that its own bond issue would be adequate, the SCEA decided to play it safe. It completed the final design and all the permitting for the 12 miles and then transferred its project to the FDOT which built the road and took over control of it.[366]

Construction of the road was divided into five sections. Hubbard Construction Company received the contract to build from S.R. 426 to S.R. 434, with toll ramps at both Red Bug Lake Road and S.R. 434. Boh Brothers contracted to build the segment from S.R. 434 to C.R. 427 which included the long bridge over Lake Jesup. There would be a toll plaza between the lake and the C.R. 427 exit. The three remaining sections from C.R. 427 to U.S. 17-92 were to be built by White Construction Company. There were no toll facilities at U.S. 17-92.[367]

The $176 million road was a sizable construction project. It required 7.4 million cubic yards of dirt and 13 miles of drainage pipe. The 1.5 mile long bridge over Lake Jesup was one of the largest inland bridges in the state and required 1,300 concrete and steel pilings ranging from 100 to 120 feet in length. Acquisition of the 644 acres of right-of-way necessitated the removal of 22 homes, seven mobile homes, one church, one business, and a 70-resident retirement center.[368]

Nor did construction proceed without difficulties. Boh Brothers finished the Lake Jesup segment about three months ahead of schedule, but both Hubbard and White encountered major delays.

Anticipating completion of the road in September 1993, officials were astounded when White Construction announced that one of its projects would not be finished until October 1, 1994. The other two sections would take until May and June. Hubbard's delays were less significant but that firm also expected delays of several months.[369]

Although White Construction made up some of its lost time, the new road opened in segments. The three mile section from S.R. 426 to Red Bug Lake Road opened on January 11, 1994. Another three miles from that point to S.R. 434 opened on April 1. The remainder of the road was ready for traffic in May 1994. Tolls were 50 cents at Red Bug, 75 cents at S.R. 434, and $1.50 at the main plaza north of Lake Jesup.[370]

Completion of the Seminole County segment in 1994 and the Southern Connector Extension two years later completed the Central Florida GreeneWay from U.S. 17-92 near Sanford on the north to I-4 just past Disney World to the southwest, a distance of about 55 miles. Only a six mile road from U.S. 17-92 westward to I-4 was needed to complete the GreeneWay.

The "Missing Link"

When it took over the original SCEA project, the FDOT agreed to support continued planning for the "missing link." With funds from the state revolving trust fund, the SCEA hired three engineering firms in 1992 to complete the final design for the six mile road. Chances for the "missing link" were improved in 1994 when the OOCEA, for environmental reasons, withdrew its support for construction of that portion of the Western Beltway across the Wekiva River between Apopka and I-4 near Sanford. Since the western portion of the I-4 Interchange was no longer needed, that reduced its cost by about $13 million. In 1995, funds from a variety of sources were assembled to build the road, and the FDOT then committed itself to the project. The SCEA would complete the design, the Central Florida district of the FDOT would acquire the right-of-way, and the Turnpike District would build it.

A ground-breaking ceremony was held on October 11, 1999, to launch construction of the first of three segments of the road. The featured speaker was U.S. Representative John L. Mica. The road will include new interchanges at C.R. 46A, Rinehart Road, and I-4. It is expected to open sometime in 2003.[371] If this schedule is met, the eastern portion of the Beltway around metropolitan Orlando will be complete.

Meanwhile, the OOCEA and other transportation agencies were already well along in the planning and construction of the Western Beltway.

Chapter 13

E-PASS: *An Electronic Toll and Traffic Management System with Automatic Vehicle Identification*

Anticipating a marked increase in traffic as the four new 1986 Project roads neared completion in 1988, OOCEA officials began discussing ways to improve toll collection operations. There had long been dissatisfaction with the existing arrangement. With the FDOT collecting its tolls, the OOCEA had no control over the costs involved or the quality of service. The equipment then in use was old, inefficient, and costly. Beyond that, it was the OOCEA which received customer complaints, but it could do nothing more than refer the matter to the FDOT's Office of Toll Operations (OTO). OOCEA officials believed that they could reduce operational costs and improve customer satisfaction if they were in control of their own collections system. One way to do that might be to contract with a private firm to collect the tolls.

In April 1988, Bob Harrell and Bill Gwynn met with representatives of the FDOT and SEACOR, a private service and management firm, to discuss the possibilities of the private sector taking over the toll collection function on all toll roads, including the Turnpike, which the FDOT operated. The FDOT agreed to study the possibility. Then, in 1989, with the consent of the Secretary of Transportation, the OOCEA initiated efforts to take over its own toll collections from the FDOT and contract with a private firm for the service.[372]

It was understood that a successful contract with a private firm would have to be based on a timely and accurate record of traffic and tolls which the existing system could not provide. Therefore, improved equipment would have to be installed and operating before the OOCEA could take over toll collection from the FDOT and implement its privatization plan.

In March 1989, Gregory Dailer, director of finance, asked the OOCEA board to approve a feasibility study "to look into new modern collection techniques." Dailer was referring to electronic toll collection. He explained that the existing equipment was not adequate to handle the increased traffic which the 1986 Project roads would bring, and that the current processing of traffic and revenue data was neither accurate nor timely.373

The labor intensive methods then in use on the expressway were more than 30 years old and quite costly. Computerized collection equipment could reduce manpower needs and concomitantly increase the number of cars which the toll lanes could handle. That in turn could reduce the need for adding more toll lanes as traffic increased. With cars spending less time idling in toll lanes, an added benefit would be less air pollution, as well as improved customer service. Most important, installation of such a system would create financial accountability which would enable the OOCEA to proceed with the privatization of its toll collection operation.374

The immediate need for an improved collection system, the desire to manage its own affairs, and the knowledge that automated systems were becoming available, led the OOCEA along a path toward an electronic toll and traffic management system known as E-PASS. E-PASS permits more than three times the number of cars per lane as the mechanical system and operates with 99.96 percent accuracy. But it was a long and arduous path because, although improved technology was available, a system of the scope and complexity envisioned by the OOCEA was not. It had to be designed and developed through a cooperative, if sometimes contentious, effort by Scientific Applications International Corporation, the prime contractor, its sub-contractor, Syntonic, Inc., the OOCEA, its general engineering consultant (PBS&J), and its sub-contractor, RCH & Associates.

The feasibility study resulted in the assembly of a team to develop the desired system. Gregory Dailer played a key role. He first worked under the direction of Executive Director Bill Gwynn, then Acting Director Joseph A. Berenis, and finally, Executive Director Harold Worrall. Project management was provided by Donald Erwin and Alex Jernigan of PBS&J. The toll collection and AVI technical consultant was Robert C. Hawkins of RCH & Associates. Hawkins and his firm developed the specifications for the proposed system.[375]

Although the specific technology was still to be developed, the OOCEA and its team envisioned the integration of an Automated Vehicle Identification (AVI) system and an Electronic Toll Collection and Traffic Management (ETTM) system which would track vehicles in the toll lanes, record payments, photograph each license number for violation enforcement, and provide a central computer with "real time" records of each transaction, as well as the operational status of the system at any given time.

The AVI component would have a transponder affixed to the front of the vehicle to communicate with an in-ground antenna as it passed through the toll lane. The proper toll would be deducted from each customer's prepaid toll account, and the gate arm would be raised in time for the vehicle to pass through without stopping.

The ETTM system would be a three-tiered computer network. Each lane would have a computer which controlled all lane components, including the AVI, the CCTV camera system, a detection system which counted the axles of each vehicle to verify the accuracy of the toll, a patron toll display visible to drivers, and, in some lanes, a computerized coin machine capable of detecting invalid coins and reporting mechanical problems. The Lane Controller would be connected to the Plaza Computer.

The Plaza Computer would receive transaction information from the lanes as well as alarms for violations, security irregularities, and equipment errors or failures. It would display to the supervisor the operational status of all traffic lanes at the main line plaza and related ramp plazas. It would consolidate lane, toll, and traffic data and send them to the Host Computer.

The Host Computer, located at the main office of the OOCEA, would collect all the information from the plazas and assemble it to provide a consolidated database for all accounting, audit, revenue, bank deposit, and other reports. The

This view of the Holland East toll plaza taken in 1998 shows the system in operation. The dedicated E-PASS lanes are white, and the availability of E-PASS in the other lanes is shown in the upper right corners of their signs. The cones are used to direct traffic to the reversible lanes during peak hours.

information would be available almost instantly and would be reliable. AVI transactions would take less than a second and each level of the system was expected to operate at 99.96 percent accuracy.[376]

In addition to the hierarchy of equipment, there were other complexities. For example, the AVI system was to be installed in all lanes, but would be exclusive only in dedicated lanes. There would be multi-functional lanes with AVI and manual collectors or AVI and automatic coin machines. Some lanes would be reversible to handle traffic in either direction at peak hours. The ETTM/AVI system was very sophisticated, and transforming it from a concept to an operating system was a huge undertaking.

The next step was to find a qualified contractor who could do the work. A Request for Proposals (RFP) for an AVI System resulted in three proposals. All were rejected as inadequate. Recognizing the gap between its desired goal and the currently available technology, the OOCEA then scheduled a workshop to which board members and staff, the vendors who had submitted proposals, and any other interested parties were invited. It was an information session to better inform everyone of what was available and what was needed. The workshop resulted in a decision to refine the specifications for the ETTM/AVI system and advertise another RFP. It was a difficult challenge because the project involved a new concept. Since it had not been done before, there was some difficulty in deciding what should be included.

The project team spent several months carefully revising and refining an RFP which was advertised in March 1991.[377]

Selecting a Contractor for the ETTM/AVI System

A committee composed of the members of the project team and several others carefully prepared the selection criteria to assure fairness in choosing a contractor. The selection was first based on the technical capabilities of the firm and then on the amount of the bids. Of the six bids received, one was eliminated for failure to comply with the RFP. Two others were eliminated on the basis of their technical qualifications. From the remaining three, the committee selected Scientific Applications International Corporation (SAIC) of San Diego, a large defense contractor which was then turning to civilian work as the nation's military needs diminished.[378] After extensive negotiations, a contract for the Toll Collection Equipment System (Project 256) was signed with SAIC in October 1991. The negotiated price was $15,379,000. Subsequent additions of the Southern Connector and the Bee Line to the original contract increased the total outlay to about $30 million.[379]

The Orlando Business Journal reported enthusiastically that plans were to have the system operating at the University Plaza on the Eastern Beltway by mid-1992. The system would then be installed on the toll lanes of the East-West Expressway, the Eastern Beltway, and the Bee Line by 1993. But those dates were far too optimistic.[380]

With its subcontractor, Syntonic, Inc., SAIC went to work on the project, and three months later, in early 1992, presented its completed version of the system, expecting to be paid. Long accustomed to working on government defense projects where performance standards were more lax and cost overruns were commonplace, SAIC had apparently viewed the contract differently from the way OOCEA officials understood it. The agency had understood the agreement as a design contract according to which the parties would work together to develop a workable product. What SAIC delivered was very far from what the OOCEA wanted.[381]

Over the next few months, the SAIC system failed each of several design reviews. In June 1992, realizing that the project was going to take some time,

SAIC officials asked for a payment schedule. In exchange for a slight reduction in the contract amount, the OOCEA agreed. Periodic payments would be made as work progressed, but the SAIC was required to provide an irrevocable letter of credit. That letter could be called if the system was not completed on time and demonstrated to be 99.96 percent accurate.[382]

Naming the System and Adding the Southern Connector

In July, in anticipation of the original schedule being met, Joe Berenis told the board that it was time to choose a name for the AVI system so that appropriate signs could be prepared and marketing plans could be made. A committee had selected five names from which the board chose E-PASS.[383]

At the same meeting, Harold Worrall, who had recently become executive director of the OOCEA, requested approval of a supplemental agreement to the SAIC contract to include the Southern Connector, which was to be completed in July 1993. SAIC/Syntonic agreed to install AVI and computerized toll equipment on all barrier plazas and ramps on the Southern Connector in time for its opening to traffic.[384]

This name was adopted in late December 1992.

Continuing Design Problems

At a July 29, 1992, briefing, the plan had been to complete the Critical Design Review and have OOCEA sign off on the design in mid-August. But the review, conducted at SAIC's San Diego facility between August 24-27, showed that more work was needed. A follow-up review with OOCEA in Orlando on September 10-11 had the same results. Another meeting at San Diego in late September showed that several changes were still necessary. These three meetings resulted in 144 items requiring attention. In late October, Dr. John Glancy of SAIC told the OOCEA board that the remaining Critical Design Review actions would be completed by November 16. OOCEA

Chairman Mandell told Glancy that if the deadline was not met, the SAIC letter of credit would be called.[385]

Harold Worrall went to San Diego on November 16 to see the software in operation. It still did not meet specifications.

At a November 1992 board meeting, OOCEA officials had a lengthy and sometimes very frank discussion with representatives from SAIC. During the ensuing weeks, there were discussions of whether to continue with SAIC or terminate the contract. Although unhappy with the continuing delays, both Harold Worrall, who was an experienced systems engineer, and Bob Mandell, who had worked closely with SAIC, had confidence in the firm's ability to perfect the system if given the time. And the alternatives were few. Termination would almost certainly result in a lengthy and expensive lawsuit, thus guaranteeing more delay. Beyond that, no other firm was any better prepared than SAIC to complete the project. As Bob Mandell put it, "there are some technical things that are going to have to be worked out." The problem was that, while the individual computers all worked well by themselves, they were still not working together at the required level of accuracy. Given the alternatives, and with these sentiments in mind, the OOCEA board agreed to stay the course and give SAIC more time to complete its work.[386] Installation of the ETTM/AVI system would be delayed for more than a year.

An SAIC test lane at its San Diego plant.

Harold Worrall (with back to camera) and Jorge Figueredo (far left) discussing equipment testing with SAIC personnel.

There was still some urgency, however. While operating equipment was in place on most of the expressway system and could be continued in use until the new equipment was ready, that was not true of the Southern Connector. Since it was scheduled for completion in July 1993, and the ETTM/AVI had been expected to be operational by that time, no other equipment had been planned for it.

SAIC Supplemental Agreement Number Four

During several weeks of negotiations, Bob Mandell, Harold Worrall, and OOCEA attorney Tom Ross reached agreement with SAIC officials on the terms of SAIC Supplemental Agreement Number Four, which was presented to the board on January 15, 1993. It was a complex document, but its major provisions gave the SAIC an extended schedule for development, production, and testing of the new equipment in return for major monetary considerations, a solution to the Southern Connector problem, and the unequivocal right of the OOCEA to terminate the contract if the revised schedule was not met. It was further agreed that Executive Director Harold Worrall would be "sole judge regarding whether or not the Build 1.5 software has been timely delivered and, after testing, is acceptable."[387]

SAIC agreed to return $8.8 million which it had received from the OOCEA in incremental payments, and to reduce the contract price by $605,000. The OOCEA would immediately receive title to all of the hardware for the Southern Connector. In addition, the SAIC would develop backup software to operate that system in the event that it failed to deliver the new software in time. And in the event the company failed to deliver the backup software, it would pay up to $200,000 for the OOCEA to purchase its own. This ensured that the agency could collect revenues on the Southern Connector from its opening day without regard to SAIC's performance.

The revised schedule called for installation and testing of the equipment in five phases. Phase I was to be installed by January 1, 1994, and the others would follow in sequence. The completion date was dependent upon test results. The new schedule extended the project about a year and a half beyond its original projected completion date.

Managing Its Own Operations and Maintenance

With the agreement of the FDOT, the OOCEA had been working since the late 1980s toward the assumption of responsibility for its own operations and maintenance. The transfer of toll collection responsibilities, dependent upon completion of the ETTM/AVI system, was in abeyance until SAIC completed its work. But in the meantime, the FDOT agreed in April 1993 to turn over maintenance of the Expressway system to the OOCEA which intended to privatize it.

Bill McKelvy became director of construction and maintenance with responsibility for oversight of a maintenance management contractor. In response to a request for proposals, four firms – ZHA International, Inc., Moreland Altobelli Associates, Inc., Highway Management Services, and GAI Consultants-Southeast, Inc. – submitted bids. After a careful selection process, Moreland Altobelli Associates was selected.

The firm was paid $610,000 for one year to manage about $4,000,000 worth of contracts for maintenance of the roadways and plazas. ZHA International replaced Moreland Altobelli after the first year.[388]

That contract did not include the ETTM/AVI system. Since its installation was completed in 1995, both its hardware and its software have been maintained by TransCore for about $2,000,000 per year. A new subsidiary of SAIC, TransCore was organized when the parent firm purchased Syntonic, Inc., and two other firms, and reorganized them under the new name to specialize in providing computerized toll systems and services throughout the nation.[389]

In anticipation of its expanded responsibilities, the OOCEA had authorized the new position of director of operations. In 1992, Harold Worrall employed Charles Gilliard, formerly of the FDOT, as operations manager. Gilliard was replaced in 1993 by David Pope, who became the first director of operations. When Pope left the position in early 1996, its functions were taken over by Jorge Figueredo who was already handling communications and marketing. Figueredo has since been director of operations, communications and marketing. He is assisted by Ron Fagan as deputy director of operations and Steve Pustelnyk as manager of communication and marketing.

OOCEA Director of Business Development and Human Resources Jacqueline D. Barr.

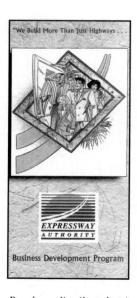

Brochure distributed to interested businesses and individuals describing the business development program, its goals, and the opportunities it offers.

A Director of Business Development

The assumption of responsibility for its own systemwide maintenance – with an annual budget of more than $5 million – enabled the OOCEA to expand its efforts to provide work opportunities for small disadvantaged and minority businesses. Instead of a few multi-million dollar construction contracts constituting most of its work program, its new maintenance responsibilities required many more services which small companies would be able to perform. A Disadvantaged Business Enterprise (DBE) committee composed of Inez Long, Robert Mandell, and Executive Director Harold Worrall recommended the appointment of a full time director of business development to strengthen the agency's Minority and Women's Business Development program.[390]

In early 1995, Jacqueline D. Barr was employed to fill that position. Under her direction the OOCEA has established relations with local minority groups such as the Minority and Women's Business Enterprise Alliance, Minority Chamber of Commerce, participated in trade fairs and workshops, and met with numerous individual businesses to publicize the agency's Equal Opportunity and Business Development program and to encourage bidding. It has also assisted interested firms in developing skills to fulfill their contracted obligations. During the five years since her appointment, Barr has succeeded in improving the OOCEA's visibility in the disadvantaged and minority business community and in increasing the number of such firms with which it does business.[391]

Completion of the ETTM/AVI System

After January 1993, SAIC kept to the revised schedule provided in Supplemental Agreement Number Four. Its backup software was installed on the Southern Connector in time for that road's opening in July 1993 albeit without the AVI component. Installation and testing of the AVI system began in early 1994 and was completed in the spring of 1995.

First made available to the public in May 1994 on the Southern Connector where traffic was lightest, the E-PASS system began attracting public interest. A lengthy article in the Orlando Sentinel in early May furnished insight into the automated system's developmental history, as well as its future promise. Referring to "a series of technical hurdles that delayed its completion," the paper reported that SAIC had corrected 1,280 problems ranging from software flaws to malfunctioning gate arms. It might have mentioned – but did not – that those problems arose while SAIC was working with 101,351 lines of code in developing a highly integrated system which had not existed anywhere before the OOCEA project began. In that context, the number of problems might have been quite reasonable.[392]

A mainline toll plaza on the Southern Connector where E-PASS was first made available.

According to the paper, Jorge Figueredo predicted that E-PASS would reduce delays at the toll plazas during rush hours and make it more convenient for expressway users who would no longer have to carry correct change. For the commuter, E-PASS would be a convenience and a time-saver.[393] Each driver could establish an account at the E-PASS service center and affix a transponder to the front of the vehicle. Then the vehicle could be driven through the lanes without stopping while the correct toll was automatically deducted from the account.

The benefit to the OOCEA was financial, the paper continued. The AVI toll lanes could handle about 1,800 cars per hour, while no more than 500 could go through

A typical day inside the E-PASS service center on Goldenrod just south of the East-West Expressway.

A toll booth operator handing a system map to a customer.

One of the flyers inserted in the Orlando Sentinel.

This E-PASS marketing piece was mailed to American Express customers giving them the option to pay their tolls through their American Express account.

the existing manual lanes during the same time. That resulted in a substantial savings because fewer new lanes would have to be added as traffic increased.[394]

News articles such as that were shortly augmented by a vigorous advertising campaign. After the equipment had been installed at several toll plazas, Jorge Figueredo began placing advertisements in the local newspapers and on the radio.

Billboards touted the advantages of E-PASS while signs posted near the toll plazas informed drivers of its availability and how to sign up. Flyers were distributed to customers when they paid their tolls. E-PASS application forms were inserted into newspapers. Local radio stations had remote broadcasts at shopping centers and commercial establishments where people could sign up for E-PASS.[395]

Having proved to be 99.96 percent accurate during the Southern Connector test, the system's installation proceeded on the remaining roadways. By October 1994, only the Curry Ford ramp plaza, the Holland East Plaza, and the Bee Line were without E-PASS. At that time there were about 5,000 subscribers. By the spring of 1995, all of the 79 miles of expressway in Orange County had E-PASS.[396]

Despite the extensive advertising program, late 1994 and early 1995 was a difficult time for the officials handling the E-PASS system. Its promise was the time saved by passing through the toll lanes without stopping, but that implied lanes dedicated to E-PASS only. Without enough E-PASS subscribers, it was impractical to provide dedicated lanes. The result was long lines with E-PASS subscribers waiting behind cars whose drivers were scrambling for change. Complaints were many.[397] Dedicated lanes could not be made available until there were enough E-PASS subscribers to make them practical, and drivers were hesitant to subscribe until they were assured that they could save time.

The History of the Orlando-Orange County Expressway Authority

WDBO radio broadcasting its Garden Rebel show from the west Orlando Target Store as new customers sign up for E-PASS.

Steve Pustelnyk and Patricia Varela making E-PASS applications available at a 1996 trade show in Orlando.

The deadlock was broken on May 1, 1995, when one E-PASS lane was opened in each direction at the Holland East and Holland West plazas of the East-West Expressway. At that time, there were more than 10,000 E-PASS users. Additional dedicated lanes were added to the system as the number of E-PASS customers increased.[398]

Removal of the Gate Arms

The ETTM/AVI system had been designed to operate with gate arms in place. Cars could go through without stopping, but they had to proceed very slowly to allow time for the gate arm to raise. At the same time, the CCTV camera system was capable of providing a photograph of every car's license plate for violation enforcement. With that capability, the OOCEA decided to remove all the gate arms, enabling E-PASS users to pass through the lanes without waiting for the gate arms to go up, although a 25-mile-per-hour limit was imposed for safety reasons. This change greatly increased the number of cars which could pass through the E-PASS lanes.

Removal of the gate arms allowed customers to pass through the lanes with much less delay.

167

Violation Enforcement

The OOCEA had long been concerned with toll cheaters and welcomed the cameras accompanying the ETTM/AVI system as an aid in reducing their numbers. The 1993 legislature empowered the agency and other toll system operators in Florida to use photographic images and impose fines on vehicles passing through the lanes without paying.[399] With the authority to act against violators, and the cameras in place by early 1995, a toll enforcement officer was employed and an enforcement policy was phased in. Realizing that it was easy for innocent drivers to make mistakes, and wishing to punish only the repeat offenders, the officer first sent warning letters with the license plate photographs to those who had committed three offenses. After the fourth offense, the offender was told to pay the toll, and a $20 fine. If there was no compliance, a ticket was issued. Costing $30 dollars in 1995, the ticket could be enforced by judges who could impose a greater fine if they so chose.[400]

By 1997, with the volume of traffic on the expressway increasing dramatically – to about 130 million transactions that year, OOCEA officials noticed a sharp increase in the number of violators. The corresponding loss of revenue was estimated at $2.36 million per year. Determined to reduce the losses, the OOCEA increased its enforcement effort with a streamlined process for warning letters, more enforcement personnel, and more tickets issued. The fine for a fourth violation had increased to $42. The enhanced enforcement policy was successful in reducing the income losses to more acceptable levels.[401]

An application brochure announcing the forthcoming opening of E-PASS lanes in Seminole County.

Expansion of E-PASS Beyond Orange County

As more and more drivers began using E-PASS, a sizable number of them lived in Seminole County and commuted to Orlando. Some of them wanted the AVI capability extended over their entire route. Gerald Brinton, executive director of the Seminole County Expressway Authority, agreed with them. "We're 100 percent behind them," he said, "It's great customer service and a real convenience for commuters."[402]

Although the FDOT, which operated the Seminole County portion of the GreeneWay (S.R. 417), was developing its own AVI system, it agreed in early 1996 to the installation of E-PASS on that road. At that time, there were more than 50,000 E-PASS customers and nearly 10,000 of them lived in Seminole County. By September 9, 1996, when E-PASS became available in Seminole County, the total number of subscribers had increased to 70,000.[403]

The agreement with FDOT also included that portion of the road in Osceola County between S.R. 535 and I-4 when it opened. Since 1995, pursuant to a contract with Osceola County, the OOCEA had been operating an AVI system known as O-PASS on the Osceola Parkway.[404]

WDBO was on air before daylight at the Target Store on Red Bug Road near Oviedo on September 7, 1996.

System Improvements

In 1996, the OOCEA began installing a fiber optic cable along the entire 79 miles of the expressway system in Orange County. Completed in 2000, it moves information through the ETTM/AVI system at the speed of light. In describing what that means, Jorge Figueredo said it would make all the computers in the entire system seem to be working as one.[405]

What began as an effort to address the so-called Y2K problem led to the recent replacement of all the ETTM/AVI computers with the latest available hardware and software. An entirely new camera system is also being installed.[406]

An Assessment

As of late 1999 the ETTM/AVI system had cost about $55 million, including $30 million for the original lease-purchase, about $10 million for maintenance since 1995, and approximately $15 million for system enhancements, including new coin baskets, Y2K adjustments, and new equipment. At the same time it saved at least $60 million by putting more vehicles through existing toll lanes instead of having to build new ones.

The Johnny Cool Band playing at a festival celebrating five years of E-PASS in 1999.

But by far the greatest benefit of all is the faster, more dependable, and efficient service to customers. Providing that kind of time-saving service has induced more and more people to establish E-PASS accounts. Ridership has grown tremendously. By 2000, there were more than 250,000 E-PASS transponders in use. Even those who do not use E-PASS benefit, because the traffic through the manual lanes is proportionately reduced. The time-saving service made possible by E-PASS has vastly increased ridership and that, in turn, has increased revenue. Tolls paid in 2000 amount to about $125,400,000.[407]

The Orlando-Orange County Expressway Authority system has become a major component of Central Florida transportation, and the ETTM/AVI system has been a major factor in making that possible.

The Five Year E-PASS celebration was held at the Goldenrod Road Service Center.

Chapter 14

Privatizing Toll Collections

Gregory Dailer's March 1989 request to "look into more modern collection techniques" was a first step toward privatizing the OOCEA's toll collection operation. It was recognized that a successful contract with a private firm would have to be based on a timely and accurate record of traffic and tolls, which the existing equipment could not provide. Therefore, the ETTM/AVI system would have to be installed and operating before the OOCEA could take over toll collections from the FDOT. It would also be necessary, as a legal matter, for the two agencies to agree in writing to make the change. The transition would also require cooperation and close coordination between them.[408]

In early 1991, Bill Gwynn received board permission to develop specifications for toll collection operations. At the same time, the position of director of operations was created and Gwynn was authorized to recruit to fill it.[409]

A supplement was made to PBS&J's general consulting contract, by which the firm with Donald Erwin as project manager, agreed to help develop a Request for Proposals (RFP) for toll operations (Project 266). RCH & Associates was employed as a subcontractor with Robert Hawkins and Phil Burkard working closely with PBS&J and OOCEA staff. In August 1991, Greg Dailer asked for approval to advertise the RFP. At that time, Hawkins explained to the board that the selection process would be the same used for the toll equipment contract.

Privatization Objectives
- Efficient toll collection operation, including reduction in current operating costs
- Sound financial accounting of revenues and assets
- Responsive courteous customer service
- Serving the best interest of Central Florida people and Authority bond holders

This and other graphics in this chapter were prepared for an oral presentation on the privatization project.

It would be divided into four parts to coincide with the ETTM/AVI project as it then existed. He laid out a schedule by which Phase One would be completed in June 1992. But there would be many delays before Project 266 reached that stage.[410]

In November 1991, the RFP was held up when the FDOT decided to draw up a statewide plan for privatizing all toll operations. Then in early 1992, the FDOT proposed that it develop and advertise an RFP for the OOCEA's privatization contract. But after extensive discussion between the two agencies, there was another change. In April, it was agreed that the OOCEA would issue its own RFP and resulting contract, and that it would draft an agreement between the two agencies for the transfer of operations responsibility. A related matter, which may have been affecting the FDOT's reluctance to relinquish control, was the status of existing FDOT employees operating OOCEA's toll facilities. Executive Director Harold Worrall told FDOT that all 275-to-300 FDOT personnel would be given first right of refusal for employment by the successful private contractor.[411] Negotiations over who was to handle the privatization RFP had consumed nearly half a year, but the delay had relatively little effect on the chronological progress of Project 266 because of the subsequent extension of the toll equipment contract with SAIC.

As in the case of the RFP for bids on the ETTM/AVI project, there was uncertainty about how the privatization RFP should be written. It was clear, however, that the document needed to state as clearly as possible what it was expected to accomplish and how that was to be done. As a result, it was developed by the OOCEA team over time with input from the vendor community.

The Project 266 RFP was advertised in July 1992 and mailed to 15 transportation consulting firms who were known to possess the qualifications and seemed likely to be interested in bidding. Among them were Morrison Knudson, Kiewit, Parsons Brinkerhoff, Martin-Marietta, Confiroute (Paris, France), and Interpark (South Africa).

When it became clear that prospective bidders needed more information, a pre-proposal conference was scheduled for August 5 and the deadline was extended. The many questions raised by interested firms were answered by Charles Gilliard, who had become operations manager on August 21. After having his responses reviewed by the FDOT, Gilliard mailed them to the interested firms. During the next few months, the Project 266 team reviewed additional questions raised by the prospective bidders, consulted with the FDOT, and refined the RFP accordingly. Bidders were informed that the deadline for responses was February 16, 1993. After all that work, Supplemental Agreement Number Four to the ETTM/AVI contract was extended for more than a year. With that, the Project 266 RFP was cancelled, to be reissued in time to coincide with the installation of the collection equipment.[412]

The Southern Connector: A Test Case

The Southern Connector was scheduled to open in July 1993, while the toll collection services would not be contracted to a private vendor for many months. It was not part of the existing toll collection agreement with the FDOT, and that agency did not have enough career service personnel to provide collectors on the new road. Accordingly, the FDOT agreed with the OOCEA's plan to contract with a private vendor for temporary personnel to collect tolls on the Southern Connector until permanent arrangements could be made. The FDOT would provide managers and supervisors. Norrell Temporary Services was awarded the contract in July 1993.

According to Wendell Lawther, who has studied and written about the privatization experience, the arrangement raised some problems. Apparently the Norrell firm was not comfortable with providing employees to be trained by the transportation agencies, preferring to provide trained employees instead. Conversely, FDOT operations personnel were unaccustomed to dealing with contract management. Perhaps the brief experience provided some idea of how to refine the forthcoming RFP, but at the very least it bridged the gap between the opening of the Southern Connector and the completion of a contract with a private vendor.[413]

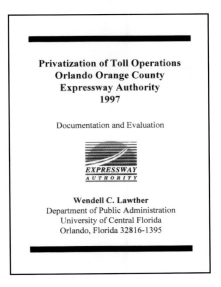

Lawther's report was published about two years after the toll collection operation was privatized.

The Second and Final RFP

When the second RFP for Project 266 was taken up in October 1993, many months had passed since the first one was canceled, and the working group had changed slightly. The OOCEA staff, including Greg Dailer, David Pope, and Jorge Figueredo, and PBS&J consultants, led by Donald Irwin, were joined by consultants from the Telos Corporation, including Phil Burkard. That firm had replaced RCH & Associates following the severe illness of its president, Robert Hawkins. The FDOT regional toll manager (Milissa Burger) was also a participant.

In its own way, the RFP for privatization of toll operations was just as complicated as the earlier one for equipment, except that it dealt with human resources, personnel management, handling money, record keeping, and public relations, instead of pioneering technology. The RFP team was much better informed than it had been a year earlier. Much had been learned from the experience of writing the first RFP. They had much better knowledge of the new toll equipment than had been the case earlier, and, perhaps, some ideas had been gleaned from observing the on-going experiment with temporary personnel on the Southern Connector. As a result OOCEA staff members and Phil Burkard of Telos Corporation suggested a number of changes, most of which dealt with increased structure and organization in the document.

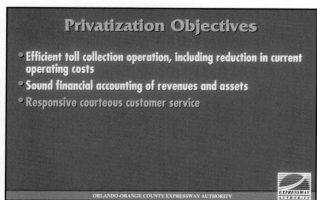

While the revised RFP stressed the desire for a partnership between the public agency and the private vendor, the resulting document was rather more specific than general.[414] A general statement asserted that privatization was intended to achieve an efficient toll collection operation, including a reduction in costs, a sound financial accounting of revenues and assets, and courteous customer service, as well as the best possible service to the OOCEA and the people of

Central Florida. It further declared that the OOCEA did not intend to replicate the current operation, but wished to provide a more creative, efficient toll operations management program.[415]

Prospective bidders were informed that operation of the toll plazas would be transferred from the FDOT to the Toll Operations Contractor (TOC) during the first year of the contract in coordination with the installation of the new ETTM/AVI equipment.

Wishing to prevent some 400 employees from being left without jobs by the transition, the OOCEA stated its desire that the contractor give FDOT career employees then working on the expressway system the first right of refusal for employment. Minimum requirements for staffing included specific mention of adequate managers, toll lane collectors, and central TOC office personnel, but flexibility was left for the vendor to implement improvements. Since employees would be handling money, the OOCEA did insist on a screening process which included a drug test and a background check for new personnel.

Although the OOCEA would train new TOC employees in the ETTM system during each phase of the transition from the FDOT, all subsequent training was the responsibility of the TOC. That training would include directional information within a 50-mile radius of Orlando as well as effective communications skills. The OOCEA and the TOC would cooperate in developing a specific customer relations procedure.

The successful bidder was to develop a mission statement aimed at achieving, among other things, a high level of customer satisfaction (with a customer complaint resolution plan), accurate collection of all tolls, and the elimination of losses of funds collected through adequate financial controls. The speed and accuracy of the ETTM/AVI system was expected to enhance the TOC's ability to achieve these goals.

The new equipment would also make it possible for the TOC to meet requirements for daily accounts of all money collected from each lane, as well as for each function involving the exchange of funds. It would further enable

A training class in progress for new personnel as the ETTM system became operational.

Manual operations continued to be important during and after implementation of the electronic system. The manual lane terminal on which instruction is being given here was central to the operation of toll booths.

A manual lane terminal in use at one of the toll booths.

an accounting of every vehicle passing through a toll lane and the resolution of money collected with the number of paid transactions recorded. With that capacity for precision, the TOC was required to record any unexplained discrepancies on a Transaction Accountability Exception Report.

There were also detailed requirements for a cost accounting system for billing purposes. This was necessary because, while the TOC would be compensated for services at a fixed rate, reimbursements for supplies would be on a cost-plus basis.

There were specific requirements for daily, weekly, and monthly reports of cash deposits, including daily reconciliations of cash deposit slips with the numbers of transactions recorded.

A toll operations standard operating procedure was to deal with hiring, training, and managing of the toll collector staff, and procedures for operating each type of toll lane. Any revenue lost because of personnel performance would be deducted from payment to the TOC.

There was also a specific statement requiring that the ETTM system be in use 24 hours a day, seven days a week so as to provide accurate processing, auditing, and reporting of every transaction. ETTM reports were to be run daily, maintained and filed in the TOC office, and made available upon request to the OOCEA. Necessary maintenance was to be coordinated with the OOCEA toll equipment contractor. System failures or malfunctions were reported automatically by a Maintenance On-Line Management System.

A section on the Vehicle Enforcement System (VES) required that the TOC maintain all video images and report violations to the OOCEA for action.

Many of the requirements were fairly specific, but the RFP included a section which listed some 35 documents, including overall plans, standard operating procedures, and reports which the TOC was required to submit after it received a notice to proceed and before it began operations. Those documents related to administration, audit/accounting, and operations. In each case, the OOCEA had the right to review and respond within 10 days, requesting changes. The apparent reason for this approach was to initiate a partnership between the OOCEA and the TOC in developing those detailed procedures which would best attain the goals sought, while permitting the agency to maintain control of its operation.416

The foregoing overview of the most significant features of the RFP demonstrates the scope of the toll collection operation and the OOCEA's determination to have an efficient, user-friendly toll system which made the best use of the new ETTM/AVI equipment. Given its comprehensive specifications, it may also show why the RFP team spent nearly five months in revising it.417

In March 1994, David Pope, who had succeeded Gilliard as director of operations, reported that the Project 266 team had held three formal meetings and many smaller ones, and their work was nearing completion. The board approved his request for permission to advertise the revised RFP in April with a view toward recommending a contractor in July or August. That would facilitate a transition from the FDOT to a private Toll Operations Contractor in October or November 1994.418

The second RFP was actually advertised on May 31, 1994. Potential bidder's site visits during June and July led to five addenda to the RFP and an extension of the due date for proposals to September 23. Bids were received from Florida Toll Services (a joint venture of Morrison Knudsen and Parsons Brinckerhoff), United Infrastructure Corporation, and URS Consultants.419

An operator reviewing and documenting toll violations.

Selecting a Toll Operations Contractor

Privatization Status

- May 94 — RFP issued
- Sept 94 — 3 proposals received
- Oct 94 — Oral presentations
- Nov 94 — FTS selected

Careful preparation of the RFP led to timely selection of Florida Toll Services for managing the toll collection operations.

Chaired by Harold Worrall who did not vote, the selection committee had five voting members. OOCEA Directors David Pope (Operations), Greg Dailer (Finance), and Bill McKelvy (Construction and Maintenance) were joined by Milissa Burger (FDOT regional toll operations manager), and Joe Kerce (Office of Toll Operations). Project Manager Donald Erwin (PBS&J) and General Counsel Tom Ross (or Lynn White) attended all meetings to provide legal or procedural advice.[420]

The bidders were required to submit detailed proposals divided into three categories, including a qualification statement, a technical proposal, and a price proposal. There was also an oral presentation. The five voting members awarded points according to a prescribed system. Their voting records were then turned over to Donald Erwin who counted the votes. On November 13, 1994, Florida Toll Services (FTS) was awarded the highest number of votes.[421] Approved by the board on November 23, 1994, a contract for the Toll Facility Operations and Management Services was signed February 28, 1995. FTS agreed to provide those services for five years for $47 million.[422]

After months of negotiations, the interlocal agreement to transfer responsibility for toll collections on the expressway system from the FDOT to the OOCEA was signed on January 25, 1995.[423]

Privatization Status

- Jan 95 — Interlocal agreement signed with FDOT
- Feb-Mar 95 — Job fair
- Mar-Sept 95 — Transitioned to FTS

After the interlocal agreement between the OOCEA and the FDOT was signed, it was only a short step to implementation of the contract with FTS.

The History of the Orlando-Orange County Expressway Authority

The Transition from FDOT to the Private Vendor

David Pope assumed responsibility for overseeing the transition process in January before the contract was signed. By February 6, he had assembled a transition team from OOCEA, FDOT, FTS, and PBS&J which met weekly through May.

A job fair was held from February 27 through March 8 to assist FDOT employees in deciding whether to remain with their present employer, transfer to FTS, or go elsewhere. Most of those with five years or more seniority quickly filled the 182 positions which remained available with the FDOT. The FTS hired 208 former FDOT employees, several of whom were promoted to fill supervisory positions. More than half of the temporary workers were also employed by the private vendor.[424]

During April and May 1995, most of the toll plazas were converted to private management without major difficulty. The last one – the airport plaza on the Bee Line – was transferred on October 1. The transition meetings ended in April, but they were continued as quality review meetings.

The transition to a private toll operations manager was a success. Although the numbers have increased as needed, in 1995 FTS was handling toll collections with a staff of 234 full-time and 100 part-time employees – more than 100 fewer than had been employed by the FDOT in 1994. Service then included 164 lanes, 10 mainline plazas, and 36 ramp plazas on 79 miles of

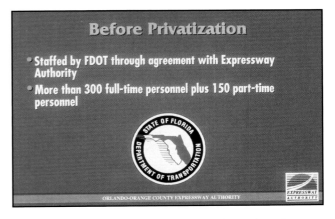

The reduction of the number of operating personnel was a major factor in reducing the cost of toll collections.

roadway. Through streamlined staffing and other improvements, about $1 million per year in operating costs was saved. Shortly after the computerized equipment was installed on the Southern Connector, David Pope showed the OOCEA board examples of information which was immediately available – information that had formerly been unavailable from the FDOT for at least two weeks. That kind of speedy and accurate information has continued and is now expected.[425]

Even more important has been the marked improvement in customer satisfaction. A customer complaint system, initiated at the outset, has shown compliments from customers rising while complaints have dropped. Although it is impossible to evaluate the two separately, the combination of more efficient, user-friendly toll collections and the time-saving features of E-PASS have vastly increased the customer base as well as the public approval of the expressway system.

The excellent performance of FTS since 1995 led the OOCEA board in 1999 to renew its contract for another five years. The new contract for $59,728,293 represented an average annual increase of 3.2 percent per year over the preceding five years. It included pay increases for toll service attendants and additional compensation and incentives for field employees, as well as annual contributions from FTS to the employee incentive plan. Noting that the quality of personnel at the toll plazas had improved over the past few years, board members felt that these personnel considerations were a good investment for the future.[426]

Chapter 15

The Western Beltway

Plans for the Western Beltway began in 1974 when Orange County financed a study of a possible route at the same time the OOCEA started a feasibility study of its eastern counterpart. That study resulted in a 1975 recommendation of an inner bypass to connect I-4 in southwestern Orange County with an extension of Maitland Boulevard. About the same time the Metropolitan Planning Organization, later renamed Metroplan Orlando, included a western outer bypass in its Orlando Urban Area Transportation Study Needs Plan (OUATS), but only the central section between S.R. 50 and Apopka was deemed financially feasible at that time.

The OUATS version of the road prevailed. By the early 1980s, plans were evolving for a 55-mile limited access road extending from an intersection with I-4 near the Osceola-Polk county line to an intersection with that road in Seminole County near S.R. 46. In discussions of a route, the approximately 30-mile section from Florida's Turnpike near Winter Garden past Apopka to I-4 near S.R. 46 was referred to as the Northwestern Beltway. It was in turn divided into two parts. Part A was an 11-mile section between the Turnpike and Apopka while Part B was a 19-mile stretch from Apopka, across the Wekiva River, to I-4. The approximately 24-mile section from the Turnpike south to I-4, originally identified as the Southwest Beltway, became Part C.

Extending through four counties – Osceola, Orange, Lake, and Seminole – the proposed Western Beltway was a large project. Not only would Part B on the north and Part C on the south pass through largely undeveloped areas making them unlikely prospects for construction as toll roads, but Part B would also have to cross the environmentally sensitive Wekiva River basin. Beyond that, both extended into other counties where the OOCEA lacked authority to build.

Part A lay entirely within Orange County in an area where the volume of vehicle traffic offered more promise of financial feasibility. Perhaps for those reasons, Part A was included in the OOCEA's 1983 <u>Long Range Expressway Plan</u>. In conjunction with the Western Extension of the Holland East-West Expressway to which it was then planned to connect, Part A was designated as the fifth of six projects the OOCEA planned to complete by the year 2000.[427]

In 1986, acting as agent for the FDOT which provided the financing, the OOCEA undertook a corridor, environmental, and alignment study of Parts A, B, and C of the Western Beltway. Although the potential road was well-received by local residents, there was some concern about the proposed route of Part A. As then planned, the road was to pass east of Apopka where it would intersect with S.R. 436. Complaints from residents on that comparatively heavily developed side of Apopka led to the selection of a more westerly route. As finally designed, Part A of the Western Beltway became a 10.5-mile road extending from the Turnpike northward between Winter Garden and Ocoee and west of Apopka to an intersection with U.S. 441. The change incidentally severed Part A from the Western Extension of the East-West Expressway which was built in the late 1980s as a part of Project 1986.[428]

The study was completed in 1988, and both the OOCEA and the FDOT adopted the final alignment for each of the three sections of the Western Beltway.[429] Already committed to completing the four roads constituting the 1986 Project, the OOCEA was in no position to finance construction of the Western Beltway at the time. But, still adhering to its role as builder of last resort, the agency looked for assistance from a new statewide road building program.

The History of the Orlando-Orange County Expressway Authority

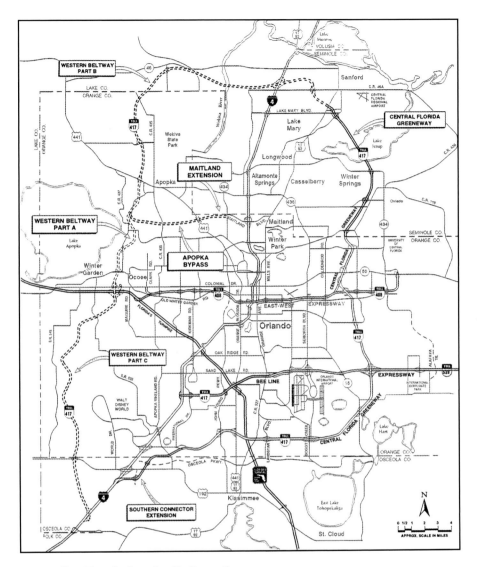

The three segments of the proposed Western Beltway are shown by the dashed line beginning at I-4 near Sanford and running west and south to I-4 in Osceola County. The Apopka Bypass is also shown in dashed lines running west from the Maitland Extension

Turnpike Funds for the Beltway?

In November 1987, Governor Bob Martinez and Transportation Secretary Kaye Henderson had toured the state touting a $40 billion, 10-year roadbuilding program. The two named the beltway around the city as one of several Central Florida transportation projects included in their program.[430] One of the ways the ambitious program would be financed was the previously mentioned "Turnpike Bill." Turnpike tolls would be increased and the proceeds would provide backing for additional construction bonds. Enacted in 1988, the legislation increased the bonding capacity of the Turnpike District by nearly $2 billion. The bonds could be used throughout the state for roads which connected with the Turnpike. To the OOCEA and other Central Florida transportation officials, the unfinished beltway around Orlando seemed an ideal candidate for such funding.

In June 1988, the OOCEA submitted the Western Beltway Part A project as a candidate for Turnpike bond funding. The FDOT went beyond that request and designated the entire Western Beltway as a part of the Turnpike system. Secretary Henderson said the road had "a high priority" for the allocation of Turnpike funds.[431] But it was the unpredictable Florida legislature which would make the final decision.

Of the 50 road projects submitted, 13 were recommended for funding. Not only did the Beltway project make the list, but it had been expanded to include the Southern Connector and the Seminole County Expressway. The FDOT was proposing to spend $931 million to build 91 miles of road to complete the Orlando Beltway. The Orlando Sentinel declared that "Completion by '94 now is possible." The editor added that "It's not a cure, but it is clear relief."[432]

Such a large appropriation would have encountered difficulty at any time, but 1989 was an especially inauspicious year. Transportation Secretary Henderson was under fire because of a huge funding discrepancy which his agency had recently reported. His boss, Republican Governor Bob Martinez, was at odds with the Democratic legislature over several issues, including the best way to finance new road construction. And a gubernatorial election was looming a year away. In the legislature, the large South Florida delegation was unwilling to support a transportation measure in which its region did not share. Other legislators were reluctant to provide large appropriations to the FDOT when it was already having difficulties with its funding.[433]

Despite all that, the Central Florida delegation, led by Representatives Daniel Webster and Tom Drage and Senator Toni Jennings, held its own. But with House Speaker Tom Gustafson and Governor Martinez locked in a battle over how to handle the transportation issue, legislative leaders decided to delay action until a special session later in the year.[434]

By the time the special session convened in November, things had changed. Secretary Henderson had resigned. New feasibility studies showed that the

Western Beltway no longer qualified for funding, but that the Polk County Parkway, in Senate President Bob Crawford's district, did. Only two segments of the Orlando Beltway, 12 miles in Seminole County and 7.7 miles in northwestern Osceola County, made the revised list of recommended projects. No legislation was enacted at that session but the FDOT's new priorities did not bode well for the Western Beltway.[435]

After another bruising battle in the 1990 legislative session, Governor Martinez finally signed a budget measure which included approval for the Turnpike to build the two segments of the Orlando Beltway mentioned above.

The Western Beltway was not included in the FDOT's five year building plan. At the same time, according to a 1990 statute, the FDOT had exclusive authority to build all three sections of the Western Beltway.[436] Further legislation would be required if the OOCEA wished to proceed with the project on its own initiative.

The Builder of Last Resort

Central Floridians were disappointed, but the OOCEA staff was not surprised at the failure of the ambitious program. The agency had continued its preliminary work on Part A while the maneuvering in Tallahassee ensued. Final design work on the road was completed in early 1990. The OOCEA's 1990 Plan Update, approved in November of that year, included the Western Beltway Part A as a "short-range need" to be constructed in the years 1996-2000 by the OOCEA. In early 1991, General Engineering Consultant PBS&J was asked to undertake a study to determine "the reasonableness of the Authority to pursue the financing and construction" of Part A.[437] But, busy with construction of the Southern Connector, development of the electronic toll collection system, and the difficulties with the city of Edgewood over the Central Connector, the OOCEA was unable immediately to take up the Western Beltway project.

Then, in November 1991, the OOCEA board voted to turn the stalled Central Connector over to the FDOT and make the Western Beltway Part A its next construction project.[438] The decision evoked widespread approval, especially

Apopka Mayor John Land is shown here addressing the audience at the ground-breaking ceremony for the Western Beltway (S.R. 429 Part A). First elected in 1949 and, except for a three-year hiatus in the late 1950's, serving continuously for nearly 50 years, the 79-year-old mayor had worked diligently with the OOCEA and other transportation authorities to have the road built.

among residents of Winter Garden, Ocoee, and Apopka. "It's the highway almost everyone wants," the Orlando Sentinel declared, but it added that, "Money is the No. 1 obstacle."[439] Both the newspaper's points were emphasized during the next three years.

With the assistance of Vollmer and Associates (traffic and earnings) and Paine Webber (finance), PBS&J completed its feasibility study – which had been expanded to include the entire Western Beltway – in 1993. It showed that, while Part A fared better than Parts B and C, projected traffic on all three parts of the road would be insufficient to make them eligible for bond funding.[440] Longtime Apopka Mayor John Land, a master of understatement, spoke for many when he said, "It's disappointing, that's for sure. We were hoping the numbers would look better than they do."[441]

In full agreement with the mayor, OOCEA officials kept the Western Beltway on their agenda, and the 1994 legislature authorized Part C of the road as a project of both the OOCEA and the FDOT. Part B remained the responsibility of the FDOT.[442]

The Apopka Bypass

With their main street jammed by through traffic on U.S. 441, Apopka residents welcomed reports in late 1992 that the FDOT was studying the possibilities of a new road which would begin at U.S. 441 near Zellwood, loop south around Apopka, and end at that highway north of Lockhart where it would connect with a planned extension of Maitland Boulevard. But in May 1994, Nancy Houston, director of the FDOT's Central Florida district, told an Apopka audience that construction of that road was at least seven years away.[443]

The Beltway 2000 Task Force

In 1994, a group of business and governmental leaders organized the Beltway 2000 Task Force aimed at having the Western Beltway built by the end of the

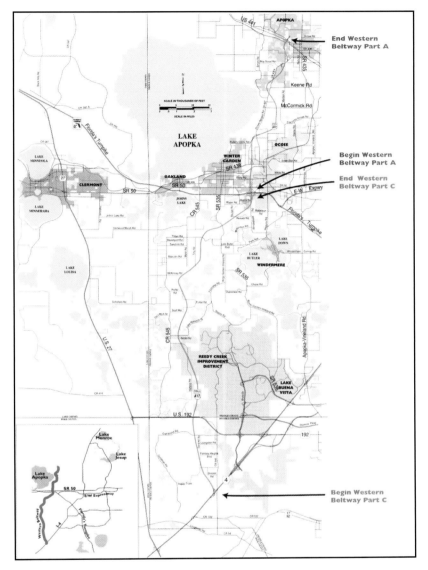

By 1994, both Parts A and Part C of the Western Beltway were designated as OOCEA projects.

century. Comprised of members from the OOCEA, the Florida Turnpike, the cities of Windermere, Winter Garden, Ocoee, and Apopka, Orange County, Walt Disney World, and firms interested in transportation, the group was specifically aimed at convincing some of the large landowners who would benefit from the new road to donate right-of-way for it.

Several members of the Task Force, County Commissioner Bob Freeman, for example, were also affiliated with Horizon West, a group of landowners then seeking changes in the county's comprehensive plan to permit more intensive development of some 60,000 acres in western Orange County adjacent to the Western Beltway corridor. Their plans were dependent upon completion of the road and, of course, their development would increase revenue projections for it.[444]

New Directions for the OOCEA

While support for the Western Beltway project was strong, the OOCEA was facing some difficult financing obstacles. Having concentrated heavily on new construction to create the existing system of 79 miles of toll roads, it had acquired a debt of more than a billion dollars. Because of the economic slowdown of the early 1990s, revenue had fallen short of projections. By the mid-1990s, costs of building a mile of expressway was about 40 times what it had been in 1965. Furthermore, the existing system required expensive improvements – addition of lanes, improved interchanges, and landscaping – to maintain a high quality of service for paying customers. These facts led OOCEA officials to the conclusion that the agency could no longer do business as it had in the past.

The OOCEA's 2015 Master Plan was based on a much different philosophy than that which had been followed in earlier years. While it would still expand its system as deemed appropriate and feasible, the OOCEA would balance its efforts between expansion and maintaining its existing system.

In September 1994, Chairman A. Wayne Rich presented a "white paper" entitled "Creating New Solutions for Central Florida's Transportation Needs" which attracted considerable attention to the OOCEA's changing role. His wide-ranging essay emphasized, among other things, that the agency lacked the financial capacity to build new roads as it had in the past. Rich proposed a partnership policy according to which the OOCEA would cooperate and share expenses with other agencies, local governments, and landowners who would benefit from new projects. For the Western Beltway he suggested Orange County, the FDOT, Osceola County, affected local governments, and representatives of adjacent landowner groups as possible partners.[445]

Approved in November 1994, the 2015 Master Plan was consistent with the chairman's "white paper." Noting that it was departing from previous practice, the plan identified four ways the OOCEA might participate in future projects. As lead partner it would accept responsibility for coordinating a variety of activities and perhaps provide financial support. As a partner, it would serve a specific role which might include a contribution of funds. As an investor it would contribute funds equal to the revenue which the new project was expected to generate. And as a contributor it would commit funds to a project with the understanding that it might not be repaid.[446]

Plans for the Western Beltway

Part A

The plan called for the OOCEA to act as lead partner for Part A with potential partners including Florida's Turnpike, Orange County, the cities involved (Winter Garden, Ocoee, and Apopka), and landowners along the right-of-way.[447]

Part C

The OOCEA was likewise to be the lead partner for Part C, also known as the Southwest Beltway. Potential partners included Orange and Osceola counties, landowners, Florida's Turnpike and equity investors.[448]

Part B

The plan called for the elimination of Part B as an OOCEA project. Instead, the agency agreed to "be an investor in the construction of the Western Beltway/Apopka Bypass interchange."[449]

This was not a surprise. As early as 1985, the agency had passed a resolution declaring that it would not support a road "in the northwest quadrant of the Orlando metropolitan area that would have an adverse environmental impact on the Wekiva River Aquatic Preserve."[450] There had been considerable discussion since then about environmental concerns and much of it had discounted the likelihood that Part B would be built.

Some considered the Apopka Bypass, which would eventually connect the Western Beltway with I-4 by way of the Maitland Boulevard Extension, as an alternative to Part B. One newspaper referred to it as a "bobtail" Beltway.[451]

But there were and still are many proponents of completing the beltway link between Apopka and I-4 in the Sanford area. While Part B was dropped from the OUATS plan as well as the OOCEA's master plan, it is still part of the FDOT's planning. In the summer of 1999, Chairman Rich listed the road as one of several projects which warranted consideration by those interested in Central Florida surface transportation.[452]

The Western Beltway Policy Committee and Task Force

To implement the new policy outlined in the 2015 Master Plan, OOCEA officials realized the necessity of including all interested agencies and organizations in the planning process. Borrowing heavily from the memberships of the Beltway 2000 Task Force and Horizon West, the agency organized the Western Beltway Policy Committee and Task Force. The Policy Committee was comprised of Orange County Commissioner Robert Freeman, Osceola County Commission Chairman Chuck Dunnick, Apopka Mayor John Land, Ocoee Mayor Scott Vandergrift, Winter Garden Mayor Jack Quesinberry, and OOCEA Chairman A. Wayne Rich. The Task Force was larger and its membership somewhat more amorphous, but Orange and Osceola counties, the cities of Apopka, Ocoee, and Winter Garden, the Reedy Creek Improvement District, the Turnpike Office, and the FDOT were always represented. The Task Force met frequently and was kept informed through progress reports on engineering, right-of-way acquisition, traffic and earnings, financing, and other matters pertaining to the Western Beltway.[453]

PD&E Study Updates

Much of the preliminary engineering work on both Parts A and C had been completed in the late 1980s, but it required updating to include any pertinent changes. Two contracts amounting to approximately $2,000,000 were let for the task. Part A was assigned to Greiner, Inc., with Mark Callahan and Drew Chapin as project leaders. Glatting, Jackson, Kercher, Anglin, Lopez & Rinehart, with Nathan Silva as project leader, was to complete the study on Part C. Beginning in May 1995, both studies were expected to be completed in one year.[454]

To keep the public informed of their progress as well as to receive "input," OOCEA Deputy Director Joe Berenis and the project leaders combined their publicity efforts. In addition to reports to the Western Beltway Policy Committee and Project Task Force, they published newsletters, met with the OOCEA's Technical Review Committee and its Environmental Advisory Group, met with homeowners associations and local governments, provided maps for viewing in their offices, and held two well-publicized public meetings.[455]

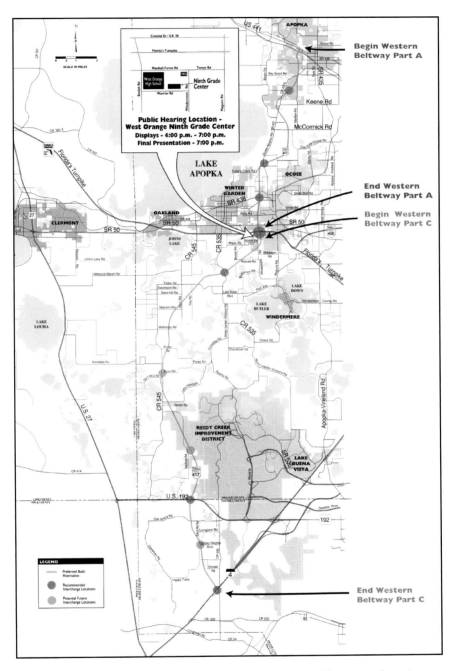

To ensure that all interested parties had the opportunity to attend, the times and locations of the meetings were widely publicized.

Joe Berenis, OOCEA deputy director, at the May 1996 meeting on the Western Beltway Part A.

Jorge Figueredo speaking at the May 1996 public forum on the Western Beltway Part A.

The two meetings, held in November 1995 and May 1996, were each attended by more than 400 people. While there was enthusiastic approval of the road at both gatherings, there were several complaints or requests for changes. Adjustments were made when possible but not everyone was satisfied.[456]

The most extensive change to the Part C design was made because the proposed alignment would have disrupted plans to create an east-west thoroughfare by connecting Roberson Road with Hartwood-Marsh Road in Winter Garden just south of the Turnpike. It would also have disrupted the planned subdivision of Stonebrook, then being developed by U. S. Homes. Lengthy negotiations with the city of Winter Garden and U. S. Homes resulted in a compromise satisfactory to all parties. U. S. Homes agreed to donate right-of-way and pay the cost of part of a frontage road. The OOCEA would pay for the remainder of the frontage road and design an interchange by which east-west traffic could pass. From additional revenue derived from the subdivision, Winter Garden would eventually reimburse the OOCEA for the cost of its part of the frontage road.[457]

A significant change in the original Part A design was derived from the FDOT's planning for the Apopka Bypass, which had begun after the 1988 design work on Part A. Since its construction was anticipated well before the Bypass was expected to be built, Greiner, Inc., was obliged to plan for the eventual intersection of the two roads. It was ultimately decided to provide an overpass on Part A south of Apopka, similar to the one which had been built on the GreeneWay in the late 1980s for the proposed Lee Vista Road.[458]

Alignments for both Parts A (June 26) and C (October 30) were adopted in 1996. About 170 parcels had been identified for Part A but fewer than a dozen houses and businesses were affected. Identification of right-of-way parcels was still in progress on Part C. Pursuant to a new right-of-way acquisition policy, land was already being acquired while the engineers were completing the alignments.[459]

The Advanced Right-of-Way Acquisition Policy

A key to the financial feasibility of the Western Beltway was finding a way to reduce the escalating cost of right-of-way. In the past, the OOCEA had waited until financing was arranged and final alignments were announced to the public before acquiring right-of-way. As Executive Director Harold Worrall put it, the bulldozers were often at work while the right-of-way was still being acquired. At the same time, the agency had relied heavily on condemnation as a method of securing the right-of-way in time for construction to begin. This approach had not only allowed land values to continue escalating while planning and engineering was being done, but it obliged the OOCEA to negotiate for land with construction deadlines looming. That policy had led to some very expensive acquisitions as well as costly administrative and legal overhead. The costly case of the Southern Connector was fresh in their minds when OOCEA officials and their consultants began exploring ways to acquire right-of-way for the Western Beltway.

Shown in the foreground is the overpass on the Western Beltway Part A for the Apopka Bypass intended eventually to connect Maitland Boulevard with U.S. 441 west of Apopka.

Comprised of Executive Director Harold Worrall, Tom Ross of Akerman, Senterfitt & Eidson, and David Brown of Broad & Cassel, a Western Beltway right-of-way planning committee went to work in early 1995.[460] Agreeing to abandon the traditional methods, the committee decided to apply a land use approach to right-of-way acquisition and to begin purchasing land before construction funding was in place. This approach would enable landowners to make the best use of their property along the new road and to use the advance purchase funds in doing so. The OOCEA would thus be better able to acquire the land at pre-development prices and by negotiation rather than litigation.

Approved in October 1995, the land use approach also offered the possibility of donated right-of-way from landowners who might benefit from the enhanced value of their property.[461]

Financing and Building Part A

By the fall of 1995, progress on the Part A project was accelerating. The 1995 legislature had appropriated $20 million for advanced right-of-way acquisition. By October, Winter Garden, Ocoee, and Apopka had approved the route of the Western Beltway within their city limits. Right-of-way acquisition began in December, nearly seven months before the final alignment was adopted.[462]

With $20 million of an estimated total cost of about $47 million in hand, the right-of-way committee was determined to make the best use of the money. Rather than scattering its purchases along the route, it decided to start at S.R. 50 and work northward in segments as funds became available. While negotiations were nearing completion on about 50 parcels representing nearly half of the needed right-of-way, the 1996 legislature appropriated

Ground-breaking for the Western Beltway Part A on June 4, 1998. Manning the shovels are (from left) Apopka Mayor John Land, Winter Garden Mayor Jack Quisenberry, State Representative Bob Sindler, A. Wayne Rich, Ocoee Mayor Scott Vandergrift, and Orange County Commissioner Tom Staley.

another $20 million for the same purpose. With that additional appropriation and a $7 million loan from the FDOT – for a total of $47 million – the OOCEA acquired the right-of-way for Part A with a minimum of litigation. On earlier projects, the OOCEA had spent about 1.85 times the appraisal price in acquiring land. Using its new acquisitions method on Part A, the agency kept its cost to about 1.2 times the appraised value.463

Final Design

Six contracts for final design of Part A were let in early 1997. They were Greiner, Inc., Projects 600 & 601; Dyer, Riddle, Mills, and Precourt, Projects 602 & 603; Bowyer-Singleton & Associates, Projects 604 & 605; Kunde-Sprecher & Associates, Project 606; and H. W. Lochner, Inc., and Kimley-Horn & Associates, Project 607. The original contracts amounted to about $11 million.464

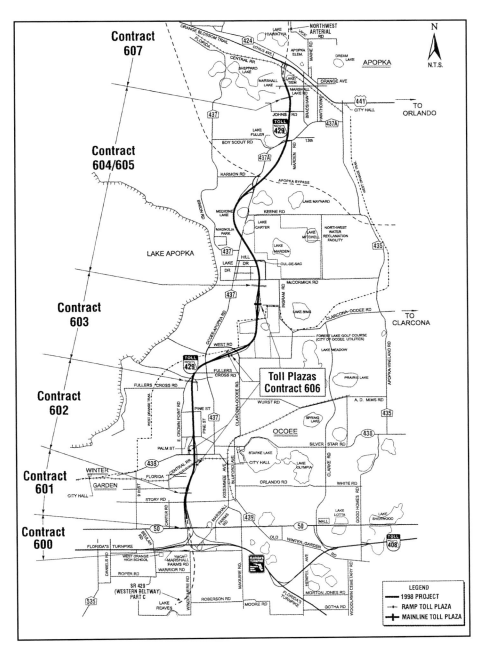

Final alignment of the Western Beltway Part A from U.S. 441 on the north to the Florida Turnpike at the south. The future continuation of Part C is shown by a dashed line.

Building A Community

S.R. 429 Part A under construction. Looking north from Marshall Farms Road past the Turnpike in the center and S.R. 50 beyond it.

The Turnpike Interchange

An interchange with the Florida Turnpike was an essential and costly element of the 10.6 mile Part A of the Western Beltway. Financing of the road was enhanced considerably when the 1995 legislature authorized the FDOT to include funding of the interchange in its next budget request.[465] Pursuant to that mandate, the FDOT included about $40.5 million in its 1998-99 work program for the interchange.[40]

Financing Part A

Due in considerable measure to the assistance from the Florida legislature, the Western Beltway Part A was becoming a financially viable project by late 1997. In February 1998, Greg Dailer stated the engineers' cost estimate of Part A at about $237 million. That included $10,800,000 which the OOCEA was spending on final design and the $47 million in grants and loans from the state for right-of-way. Mitigation for damage to wet lands was about $6,700,000. Toll equipment amounted to about $4 million. Anticipated cost of construction was $168,000,000, $40.5 million of which was being provided by the Turnpike District for construction of the interchange.[467]

Anticipated toll revenue, with the customary pledge of Orange County gasoline tax revenue, would support a bond issue sufficient to provide the approximately $129.5 million needed to complete the project. Accordingly, with Paine Webber, Inc., acting as senior underwriter, a bond issue of $200 million, some of which was to be used for other purposes, was sold in May 1998.[468]

Senator Daniel Webster (right) with Mel Martinez, Orange County chairman at the S.R. 429 Part A Grand Opening.

Building the Road

The OOCEA's methods of selecting contractors and managing construction for Part A also differed from previous practices.

Lawsuits emanating from Hubbard Construction's claims on the Southern Connector projects had resulted in years of expensive litigation and the agency was determined to prevent such problems in the future. After careful study, it developed a dispute resolution and binding arbitration policy by which contractors would have to agree to abide. Accordingly, a dispute resolution committee, composed of an appointee of the contractor, another of the OOCEA, and a third selected by the other two members, would attempt to resolve disputes. If that failed, the matter would go to binding arbitration. The procedure was included in all Part A construction contracts.[469]

S.R. 429 Part A construction at S.R. 50 on August 10, 1999.

For earlier projects, the OOCEA had contracted separately with a construction management consultant and CE&I firms. In order to provide a more direct line of communication, it was decided to hire the management consultant who would, in turn, contract with the CE&I firms. Parsons, Brinckerhoff Construction Services was selected as construction management consultant for Part A. The contract was for $20,355,468.79.[470]

S.R. 429 Part A construction at Johns Road bridge on August 10, 1999.

View south across the mainline toll plaza and the West Orange Trail. Note the new design of the toll plaza which allows E-PASS users to continue unimpeded while other vehicles veer off to pay tolls.

View south across S.R. 441 looking at the future alignment of S.R. 429 Part A.

Two contracts for concrete piles, beams, and walls amounted to $4.9 million.[471]

Construction was launched with a ground-breaking ceremony on June 4, 1998. There were six roadway contracts and one for toll plaza construction. The 10.6 mile road would have 41 bridges and ramps, and six interchanges. There would be a mainline toll plaza and five ramp plazas. The toll at the mainline plaza would be one dollar. Ramp tolls would be twenty-five cents and fifty cents.[472]

Project 601, from S.R. 50 to Story Road, was let to Hubbard Construction Company for $8,284,829. Project 602, from Story Road to Fullers Crossroad, also went to Hubbard. The amount was $16,647,283. Project 603, from Fullers Crossroad to McCormick Road, was a third Hubbard project. That contract amounted to $19,875,000.[473]

Looking north from the Florida Central Railroad Line past S.R. 438 and Palm Drive.

The successful bid for Project 604, from McCormick Road to Johns Road, came from Granite Construction Company. The contract was for $16,171,000. Project 607, the U.S. 441 Interchange, was also let to Granite Construction Company for $13,013,000.[474]

Project 600, the Turnpike Interchange, became Granite Construction Company's third Part A project at the contracted price of $32,776,000. Project 606, for the toll plazas, was let to Southland Construction, Inc., for $6,339,000.[475]

Work progressed without interruption and the road from U.S. 441 south to Winter Garden opened on July 8, 2000.

Looking south from the U.S. 441 intersection to Johns Road.

Part C and the Public/Private Partnership

The approximately 22-mile Part C, or Southwest Beltway, was a more complex project than Part A had been. Extending through a sparsely developed area, it fared poorly in traffic and earnings studies. At the same time, the portion between U.S. 192 and I-4 was rapidly filling up with commercial establishments, making right-of-way more and more expensive. The road's corridor extended through two counties. The OOCEA lacked the funding capacity to build the road without substantial assistance.

But despite these obstacles many people were interested in having the road built.

The dedication ceremony for S.R. 429 Part A on July 8, 2000 was well attended.

Building A Community

James Pugh, OOCEA chairman, addressing the audience at the opening ceremony for S.R. 429 Part A.

OOCEA board members assembled at the dedication ceremony for S.R. 429 Part A. They are (from left) Mike Snyder, FDOT, ex officio; Mel Martinez, Orange County chairman, ex officio; A. Wayne Rich; James Pugh; and Inez Long.

It was an excellent candidate for a public/private partnership as outlined in Chairman Rich's "white paper" and the 2015 Master Plan. The 1995 legislature had amended existing legislation to authorize partnerships for the Western Beltway. Another amendment permitted the OOCEA to acquire land by condemnation in Osceola County.[476]

Horizon West

By the time the Glatting Jackson PD&E study update was completed and the Part C alignment was adopted in October 1996, the OOCEA had already been cooperating with Orange County and Horizon West. Their cooperative goal was to transform the former citrus land in the southwest portion of the county into well-developed neighborhoods which would benefit the landowners and developers as well as the proponents of the Western Beltway. Together, they obtained state support for a change of Orange County's comprehensive plan to permit more intensive development of the area. But, it would be well-planned development. The Village Land Use Classification envisioned several neighborhoods designed in such a way that all residents would be within short distances of schools, parks, shopping facilities, and other amenities. Only those trips of some distance, to work, for example, would put cars on arterial roads and expressways. The development would, in turn, increase ridership on the Western Beltway. And since the road was essential to the development, there was the possibility of land donations for right-of-way.[477]

The Search for Funds

The OOCEA continued its engineering work on Part C while exploring ways to finance it.

Figured in 1998 dollars, the estimated cost of Part C was about $360 million, more than twice what was projected to be available. During 1997 and 1998, Chairman Rich and Executive Director Worrall had many conferences with Turnpike District and FDOT officials, state legislators, and private landowners, exploring various ways of building the road. They considered building it southward and stopping at U.S. 192, thus avoiding having to buy expensive right-of-way along a portion of the road expected to be lightly travelled. There were discussions of making that part of the road "controlled access," which was less expensive than "limited access." Perhaps more important in the long run were considerations of dividing the project, with the OOCEA building the northern portion while the Turnpike District built the remainder.

Of immediate concern to OOCEA officials was the necessity of acquiring the right-of-way before development placed it beyond reach. As of April 1996, right-of-way costs were estimated at $88 million. Fifty-three million of that was located along the small portion of the road between the Reedy Creek Improvement District and I-4.[478] In order to avoid increasing costs where development was imminent, the OOCEA had already spent about $16 million on property just south of the Turnpike and at Orange Lake Country Club.[479]

As had been the case with Part A, the Central Florida legislative delegation, especially Senate President Toni Jennings and House Speaker Daniel Webster, was working hard to secure funding for the road. Its efforts resulted in a 1998 appropriation of $60 million for advanced right-of-way purchases. Meanwhile, Webster was also meeting with FDOT, Disney, Horizon West, and other landowners along the corridor to develop a financing plan.[480]

Webster's interest stimulated efforts already underway by the OOCEA. As lead partner for the Part C project, the agency had already made considerable progress toward a public/private partnership. With strong interest in an

Building A Community

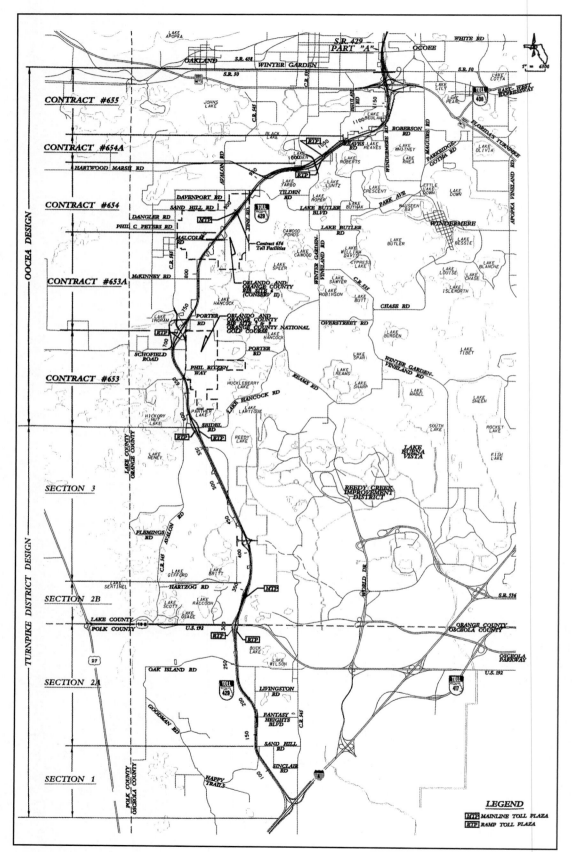

Final alignment of the Western Beltway Part C (now S.R. 429 Part A). Note the division at Seidel Road. Pursuant to a partnering arrangement, the OOCEA will build from the Turnpike to Seidel Road, and the Turnpike District will build from that point through the Reedy Creek Improvement District property and past the Osceola County line to I-4.

expressway which would provide access to their west side, Disney officials agreed to provide about $4 million worth of right-of-way, mostly through the Reedy Creek Improvement District, and $7.5 million in cash. The Horizon West group was unwilling to make a firm commitment, but would encourage the donation or sale of right-of-way at pre-development prices.

At least one OOCEA board member thought the private owners' concessions were inadequate when compared to the benefits they were receiving, and Disney's proposed contribution came with onerous conditions. But despite such reservations, and with considerable encouragement from the 1998 legislature, an understanding was reached in April 1998 by which the FDOT, the OOCEA, Walt Disney World Co., the Reedy Creek Improvement District, Lake Buena Vista Communities, Inc., and Horizon West agreed to negotiate in good faith toward a Joint Partnership Agreement (JPA) for finalizing the planning and financing (but not construction) of Part C.[481]

JPA 1998, with Orange and Osceola counties added as partners, was signed in February 1999. Perhaps its most salient feature was the division of the Part C project into two approximately equal parts. The OOCEA would build about 11 miles from the Turnpike south to Seidel Road while the Turnpike District would build the remainder from that point to I-4. Each would proceed with design and right-of-way acquisition in its respective portion. The OOCEA would be reimbursed $13.4 million for its purchase of the Orange Lake and Falin parcels (south of Seidel Road) to be used for right-of-way acquisition in its portion of Part C. The earlier commitments by the Disney companies and Horizon West were restated. The OOCEA committed to spending $32 million on the project. Sixteen million of that had already been spent for right-of-way as had $14 million for design. Another $2 million would go for final design. The parties agreed to work toward completion of JPA 2000 by October 15, 2000. That agreement would address the financing, construction, ownership, and operation of Part C of the Western Beltway.[482]

In early 2000 an agreement was reached whereby the FDOT would grant $15 million to the OOCEA toward construction of Part C from the Turnpike to Seidel Road, and pay for its operations and maintenance from 2003 to 2005. That would enable the OOCEA to build the road from the Turnpike to S.R. 535 by 2003 and from that point to Seidel Road by 2005. The Turnpike District agreed to build from Seidel Road to U.S. 192 by 2005 and to I-4 by 2008.[483]

A New Name for the Western Beltway

With Part B no longer an OOCEA project and prospects for its construction in doubt, OOCEA staff members pointed out that the road was no longer a beltway. Accordingly, its name was changed to Western Expressway. The change was more symbolic than substantive, however, since Parts A and C, when completed, will be known officially as Toll 429.[484]

Chapter 16

The OOCEA in 2000 and Beyond

The new directions adopted in the 2015 Master Plan were in place by 2000. While construction of the Western Expressway Part A was progressing and plans for Part C were being finalized, the OOCEA was working on an approximately $500 million system improvement project to widen the Holland East-West Expressway (Toll 408) as well as a massive reconstruction of that road's interchange with I-4. System improvements were in balance with system expansion. The agency's new policy of building roads through public/private partnerships was meeting with greater success in southeast Orlando than was the case of the Western Beltway Part C. It was also addressing the necessity of adapting its successful E-PASS system to FDOT requirements and working on its 2025 Master Plan and several of the projects which it would include.

Widening the East-West Expressway

With the continuing growth of metropolitan Orlando, the expansion of the expressway system, and the success of E-PASS, the East-West Expressway from its intersection with the Central Florida GreeneWay (Toll 417) on the east, through downtown, to Kirkman Road on the west was becoming quite crowded. Backups were occurring during the morning and afternoon rush hours. An early 1998 traffic analysis by PBS&J, Leisch & Associates, and Transportation Consulting Group projected traffic volumes soon exceeding acceptable levels on portions of the road from I-4 to the GreeneWay.[485]

Studies of the road west of I-4 to Kirkman Road in 1997 and east from I-4 to the GreeneWay in 1999 evolved into plans for an approximately $500 million project which would expand the road to three lanes west of downtown and four lanes east of downtown. It would include widening of the Lake Underhill bridge, as well as adding noise walls at appropriate points. Construction is anticipated to begin in 2003.[486]

For planning purposes, the widening project was divided into three segments: from Kirkman Road to John Young Parkway, from Tampa Avenue to I-4, and from I-4 to the GreeneWay. A correlative project, from U.S. 441 east to Rosalind Avenue, is the East-West/I-4 Interchange.[487]

The East-West/I-4 Interchange

The need for improving the East-West Expressway interchange with I-4 had become apparent in the early 1990s. The westbound exit ramp from the East-West Expressway to I-4 was especially overtaxed. By 1997, it was handling 2,400 vehicles per hour during peak periods when it was designed to carry only 1,800. Resulting backups extended past Rosalind Avenue, creating operational as well as safety problems.[488]

A complex interchange had been designed to alleviate the problem, but there were obstacles to its construction. Not only was its estimated cost of $350 million beyond the means of the OOCEA or the FDOT, its implementation would disrupt other thoroughfares as well as several neighborhoods. Alternative plans designed to address these matters not only increased the cost, but added complexity and difficulty of operation to the proposed project. A further complication was that any

Conceptual design of a new interchange at the intersection of Toll 408 and I-4.

construction involving I-4 had to be approved by the Federal Highway Administration. Ongoing plans by the FDOT to widen I-4 also required coordination.[489]

While plans for the full interchange are still being considered by several agencies, the OOCEA has retreated to a possible interim solution involving ramps connecting the East-West Expressway westbound to I-4 eastbound and I-4 eastbound to the East-West Expressway eastbound. The FDOT has agreed to a partnership by which the two agencies will share the costs of this interim project. It is included in the current Five Year Work Plan and is planned for completion in 2007. The FDOT and the OOCEA will each pay half of the estimated $35 million for right-of-way. Depending upon the ultimate cost of the project, the OOCEA may have to issue bonds to fund its portion of the construction price. Completion of the East-West Expressway widening project is dependent upon resolution of the interchange matter since, as Executive Director Harold Worrall put it, "these two projects directly interrelate."[490]

The Goldenrod Extension and the Lee Vista Interchange

The need for additional roads in southeast Orlando and the inability of local governments to provide them were the catalysts for the successful implementation of a public/private partnership in which the OOCEA acted as lead partner as well as a contributor.[491]

Concern about the phenomenal growth of Orlando International Airport and rapid development of the area to its east stimulated the 1995 Southeast Orange Transportation Analysis (SOTA) in which the OOCEA participated along with the FDOT, the GOAA, Orange County, the city of Orlando, and Lynx. Completed in 1997, the analysis concluded that roads in the area were inadequate to handle anticipated traffic. The developments of Lee Vista and Lake Nona were well underway and others were on the drawing boards. All participants in the study agreed that another access to the burgeoning airport was essential. While the study was proceeding some of the landowners who had originally negotiated to have an interchange on the GreeneWay at a

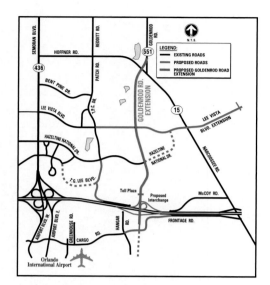

Map showing the several projects associated with the Goldenrod Road extension. The Goldenrod Road extension from Hoffner to south of the Bee Line is shown by the heavy black line. Just south of the Bee Line interchange, Cargo Road will be built to the airport. Lee Vista Boulevard will be extended to S.R. 417 as shown. Construction has begun on the Lee Vista/S.R. 417 interchange.

Looking north at the off-ramp from S.R. 417 south to Lee Vista Road. The main line toll plaza is visible in the distance.

proposed Lee Vista Boulevard approached the OOCEA about completing that long-delayed project.[492]

Because resources were lacking to fulfill these needs, the OOCEA offered to become lead partner in a public/private partnership to extend Goldenrod Road from its existing intersection with Hoffner Road (S.R. 15) about two miles through an intersection with the Bee Line, to an extension of Cargo Road which the GOAA would build. It would also complete the GreeneWay/Lee Vista interchange which had been left unfinished in 1988. The city of Orlando and the adjacent landowners would build Lee Vista Boulevard which had been planned but not built in 1988.[493]

Plans for the Goldenrod Road Extension called for a four-lane, controlled access road from Hoffner Road to the Bee Line where an interchange would be built. There would be a toll plaza just to the north of the Bee Line where the toll would be 50 cents. Goldenrod Road would then connect on the south side of the Bee Line with a new portion of Cargo Road leading into the airport. The overall cost of the project was estimated at about $30,875,000. Lengthy negotiations between the OOCEA, the GOAA, the city of Orlando, Orange County, and the several landowners, resulted in a complex arrangement, spelled out in three separate agreements, by which the road would be financed and built. The city agreed to contribute $2,000,000, the GOAA $4,500,000, and the county $1,000,000. The OOCEA would provide the remaining $23,375,000. All parties would be reimbursed from toll receipts. When they were repaid in an estimated 15-16 years, the toll plaza would be eliminated and the city would assume ownership of the road while the interchange would belong to the OOCEA. Since the OOCEA would be repaid only the principle, it was a contributor to the project in the amount of the interest it would lose.[494]

In a related agreement, the OOCEA, the city of Orlando, and the property owners who would benefit, agreed to build the Lee Vista Boulevard Interchange at the GreeneWay. The property owners would contribute the right-of-way and $800,000 while the city would pay $1,100,000. The OOCEA would provide the remaining amount of approximately $3,600,000. The city and the property owners would build the Lee Vista Boulevard Extension as well as an extension of Chickasaw Trail to connect with it.[495]

As Chairman Rich put it at a February 1999 meeting, completing the public/private partnership had been a "long process and a very complex transaction" involving a "host of entities." He complimented everyone involved, including the agency's legal team of Broad & Cassel which had been led by David Brown.[496] Rich had good reason to be pleased. The Goldenrod Agreement was exactly the kind of partnership envisioned in his 1994 "white paper" and the 2015 Master Plan.

Permitting, mitigation, and right-of-way acquisition was completed in 2000. Construction began in 2001.[497]

SunPass

When the OOCEA began planning its highly successful E-PASS system and the privatization of toll operations, the FDOT was also working toward such a system for the entire state. The two agencies initially agreed to coordinate their efforts, but the FDOT was delayed and the OOCEA was permitted to proceed on its own because its existing toll facilities were facing significant traffic congestion. As has been shown elsewhere, the E-PASS system was installed in 1994-95 and was soon contributing to the rapid increase of traffic on the expressway system.

The OOCEA was told in late 1997 that it would soon be required to convert its electronic toll system from its E-PASS to the FDOT's state-wide SunPass. In early 1999, it was further learned that the FDOT's Office of Toll Operations envisioned the transition as a simple matter of switching from one system to the other. OOCEA Director of Operations Jorge Figueredo felt strongly that the two systems needed to remain independent but ultimately compatible. Eager to provide uninterrupted service for the agency's large number of customers, Figueredo sought a longer transition period during which the two systems would operate together while the existing bumper-mounted transponders were gradually phased out in favor of windshield-mounted units compatible with SunPass. The technology for this procedure was developed, but the state agency mandated a shorter transition time.[498]

Another matter of concern was the accuracy of the SunPass equipment. While the FDOT's contract with the equipment provider calls for 99.97 percent accuracy, it was operating at only 95 percent when installed in South Florida in early 1999. E-PASS had consistently operated at 99.96 percent accuracy.[499]

While the FDOT Office of Toll Operations continued to work toward improving its equipment, the achievement of SunPass compatibility was accomplished by mid-2001.[500]

Intelligent Transportation Systems

E-PASS and SunPass are examples of a new genre of equipment known collectively as Intelligent Transportation Systems. Included are systems which provide traffic advice on billboards, speed sensors, and information to be made available to appropriately equipped vehicles. Region-wide traffic management systems are also available. With its fiber optic cable network in place, the OOCEA is capable of implementing those ITS systems in the future. In addition, the agency also plans to lease its excess space to interested commercial firms.

Completion of the fiber optic cable on the expressway system provides the technology for informational signs such as this one on I-4.

This technician is installing fiber optics.

Installation of the fiber optic cable was a large undertaking. Here the cable is being buried along the expressway right-of-way.

Expansion of the OOCEA's Scope

In the early 1970s, the OOCEA's effort to expand its expressway system into Seminole and Osceola counties had met strong suspicion and resistance. Expansion of the system and its demonstrated utility in relieving the Central Florida region's traffic congestion transformed the agency from a suspected intruder to a welcome partner. Adjacent counties were increasingly willing to permit OOCEA roads within their boundaries. Transportation officials noted that the expressway system did not stop at county boundaries. And customers, concerned with destinations, paid little heed to the county in which they were driving. With metropolitan Orlando continuing to spread across Central Florida, transportation needs necessarily reached into other counties. To keep up with changing needs, the 1997 legislature changed the OOCEA charter to permit the agency to make interlocal agreements to build and operate roads in other counties.[501]

Extensions into Lake County

By 1997, population growth in southeastern Lake County was causing increasing traffic from there into central Orange County. This was first made clear when Tom Biggs of Transportation Consulting Group was conducting an Origins and Destination Study regarding potential travel on the Western Expressway Part A. He found "very significant travel patterns…through the U.S. 441 corridor…from Lake County into central Orange County."[502] That led

William A. Beckett, OOCEA member, 1995-1999.

Upon his election as Orange County chairman, Mel Martinez succeeded Linda Chapin as ex officio member of the OOCEA.

to further investigation of a possible northwestern extension from the Western Expressway Part A near Apopka to some point in Lake County near Mount Dora. PBS&J personnel conducted a study and reported four possible corridors together with estimated costs in November 1999. Executive Director Hal Worrall, Joe Berenis, and other staff members met with Lake County officials who were interested in pursuing such a project.[503]

By that time, it was becoming apparent that traffic was also growing farther south along the S.R. 50 corridor. Morning rush hour traffic from Lake County to the Orlando area was jamming the Turnpike toll plaza at S.R. 50. By 1999, S.R. 50 traffic was backing up for a mile waiting to enter the Turnpike.[504] Discussions with Lake County staff members revealed considerable interest in a road extension in that area as well. The discussion centered on the possibility of extending the East-West Expressway westward to an intersection with U.S. 27.[505] In May 2000 the Lake County Commission approved a interlocal agreement to allow the Expressway Authority to study possible routes into Lake County.

Personnel Changes, 1995-2000

Several changes in OOCEA board membership occurred between 1995 and 2000. In 1995 Robert Mandell's term ended and he was succeeded by William A. Beckett. Beckett was then replaced in 1999 by James H. Pugh, Jr. When Wayne Rich stepped down as chairman in late 1999, Pugh was elected to that position. Succeeding Linda Chapin as Orange County chairman in 1998, Mel Martinez became the ex officio member representing the county. In late 1999, Nancy Houston was replaced by Mike Snyder as the ex officio member representing the FDOT.

Since Harold Worrall became executive director in 1992, there have been only three major personnel changes in the OOCEA's management team. Jacqueline D. Barr became director of business development in 1995. Charles Sylvester retired in 1998. Chief Financial Officer Gregory Dailer was replaced in 1999 by Teresa Slack.

Past, Present and Future

The OOCEA has come a long way in the 37 years since its inception in 1963. Charged with the responsibility of providing toll roads as supplements to the transportation needs of more than 300,000 inhabitants of the metropolitan Orlando area, the OOCEA was managed in its early years by a few public-spirited volunteers. Their only professional engineering advice came from Charlie Sylvester, the FDOT's liaison officer to the OOCEA, and contract engineers such as Harry Bertossa of the firm of Howard, Needles, Tammen, and Bergendorff. In this fashion, Richard Fletcher, James Greene, and their colleagues built the original segments of the Bee Line and the East-West Expressway between 1965 and 1973.

But, before any earth was moved on either of those projects, new developments associated with the space program and Walt Disney World set in motion a period of growth which has increased the 300,000 plus population of the metropolitan Orlando area in the 1960s to more than one and a half million in 2000. In cooperation with local, regional and state transportation agencies, the OOCEA has been busy ever since trying to keep up with that growth.

In the process, it has gone from an agency comprised of volunteer board members who contracted with professional firms to build roads which were operated and maintained by the FDOT, to an agency with an experienced diverse professional staff. It is assisted by a general engineering consultant capable of long range planning and the management of numerous contracts dealing with construction, maintenance, and operation of its 90 miles of expressways. In the process it has kept up with technological improvements around the country and actually advanced the field of intelligent transportation with the development of its innovative ETTM/AVI (E-PASS) system.

As a contributor to the transportation network of the metropolitan Orlando area, the OOCEA's expressway system can be measured in many ways. For the

James H. Pugh, Jr. filled the vacancy left by William A. Beckett and then succeeded A. Wayne Rich as OOCEA chairman.

Mike Snyder replaced Nancy Houston as ex officio member from the FDOT in 1999.

View north across Lake Lucerne and the Orlando skyline in the late 1990s. Both the north shore of the lake and downtown Orlando have undergone extensive changes since the pictures of the 1950s shown in Chapter 1 were taken.

individual who uses the system, it is a safer way to get from one place to another in a shorter time. For drivers who do not use it, the system reduces the volume of traffic on other roads.

For the metropolitan area, downtown Orlando, the University/Research Park area, Orlando International Airport, the southwest Orange County theme park area, and Maitland/Altamonte Springs have been identified as the five major trip generating centers. The expressway system serves four of them with only Maitland/Altamonte Springs being excepted.[506]

A 1997 study by the University of South Florida's Center for Urban Transportation Research offered a different measurement. According to the Center, without the expressway system, area drivers in 1993 would have spent an extra 28,477 hours in traffic every day, using an extra 18,478 gallons of fuel, amounting to 6.7 million gallons per year. For that year, the dollar value of time saved was estimated at $150 million, the increased safety of the

A different kind of measure of the expressway system was made in 1998 when the Central Florida GreeneWay (Toll 417) was judged by the American Automobile Association to be "One of America's Top Ten Roads."

expressway system was worth $73 million, and the fuel cost savings was $5 million for an overall saving of $228 million. Projecting its estimate for 1993 to 2020, the Center suggested the system's overall impact at between $4.1 and $5.4 billion.[507] Increasing traffic since the study was completed reinforces that projection.

As of 2000, the OOCEA had an annual income of $125 million from toll revenues alone. It spends approximately $30 million per year on operations and maintenance while managing a debt of well over a billion dollars. With the completion of the Western Expressway Part A, it will have built five of the six construction projects included in its 1983 Long Range Expressway Plan. With the addition of that road, the expressway system mileage is nearly 90 miles. The impending completion of the OOCEA's portion of Part C will increase its system to slightly more than 100 miles.

The expressway system as it existed in 2000. The Western Expressway (S.R. 429 Part C), shown here in dashed lines, is in the planning stages.

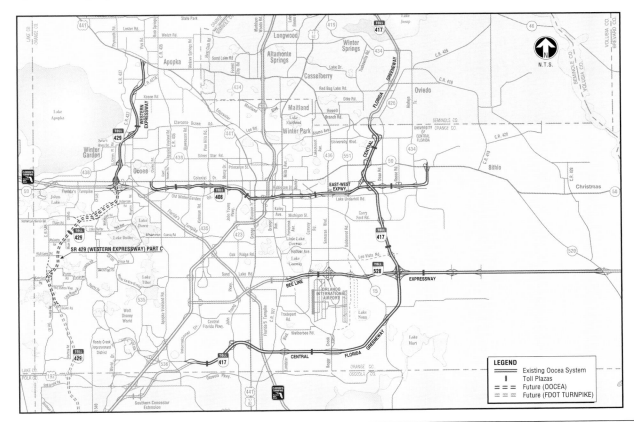

Building A Community

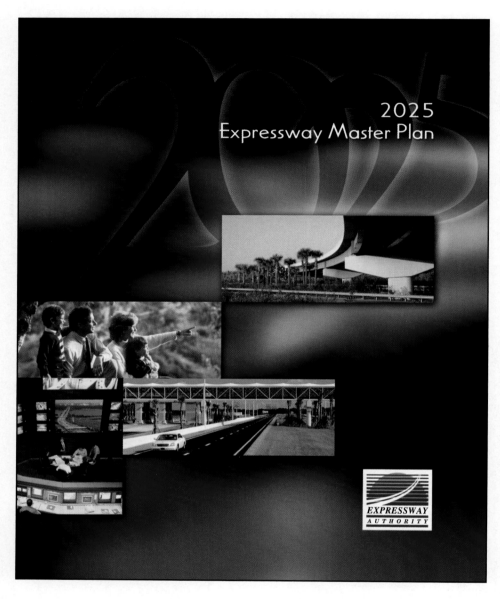

The 2025 Long Range Master Plan

In the future, using its new partnering approach, the agency can be expected to add other roads while continuing to improve and maintain its existing system. Adopted in late 2000, the 2025 Long Range Master Plan envisions a program of system improvements including the addition of lanes, ramps, and interchanges to keep up with projected traffic growth. It also envisions the agencies participation as lead partner in four proposed expansion projects and as a partner in three others.

One of the four projects in which the OOCEA will be lead partner is the previously discussed Western Expressway Part C. Another is S.R. 408 (East-West Expressway) Western Extension which is envisioned as a reliever to S.R. 50 extending from Ocoee west to U.S. 27 near Clermont. S.R. 429 (Western Expressway) Northwest Extension is a proposed new project to extend S.R. 429 Part A from U.S. 441 north and west toward Lake County as a reliever to U.S. 441 in northwest

The History of the Orlando-Orange County Expressway Authority

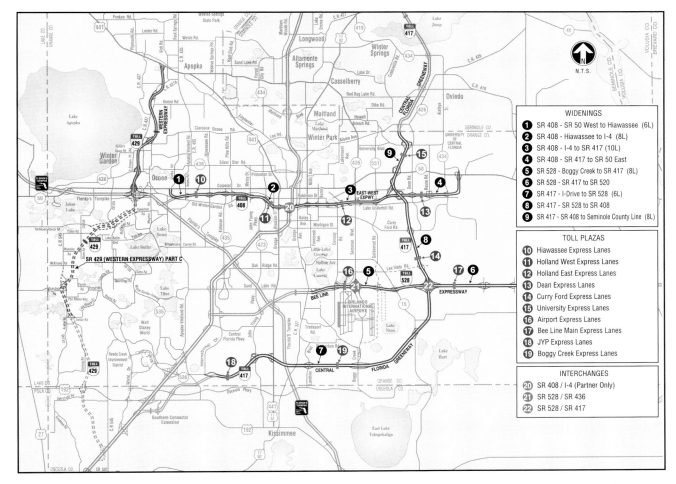

2025 Master Plan proposed Expressway improvements.

Orange County. S.R. 408 (East-West Expressway) Eastern Extension is a proposed addition to S.R. 408 from its current terminus to S.R. 520 as a possible reliever to S.R. 50 in east Orange County. Future extensions could continue east toward I-95.[508]

The OOCEA could be a partner in the construction of a new toll road from the GreeneWay in Seminole County north into Volusia County near Deltona to serve as an alternate to I-4. S.R. 417 (GreeneWay). On S.R. 436, plans include the construction of an elevated limited access roadway equiped with High Occupancy Toll (HOT) lanes above S.R. 436. Extending from S.R. 408 (East-West Expressway) to S.R. 528 (Bee Line), it would provide improved access to Orlando International Airport.[509]

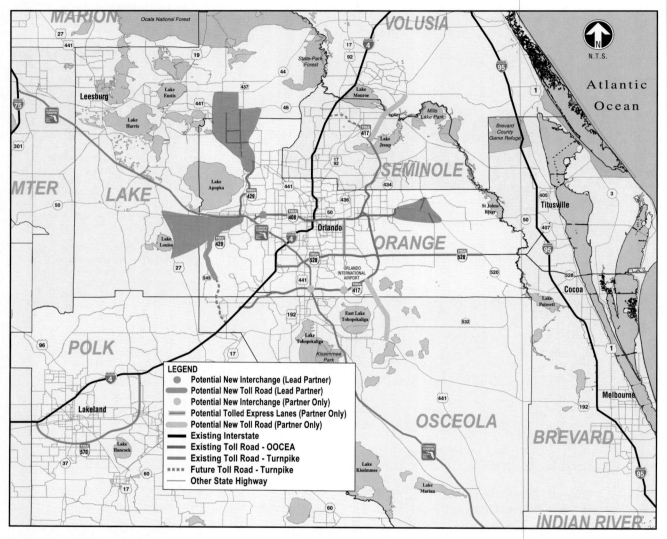

2025 Master Plan proposed Expressway expansions.

The OOCEA envisions its role in the last three projects as a partner because the first two are located in adjacent counties and the third would require considerable contributions from other partners. But, as an integral component of the Central Florida transportation community, the OOCEA can be expected to continue to cooperate with other agencies to cope with the expanding needs for more and better ways of getting people to their destinations and bringing them home.

List of Abbreviations

AVI	Automated Vehicle Identification
CCTV	Closed Circuit Television
CEI	Construction, Engineering, and Inspection
CFDC	Central Florida Development Commission
DBE	Disadvantaged Business Enterprise
DER	Department of Environmental Regulation
ECFRPC	East Central Florida Regional Planning Council
ETTM	Electronic Toll and Traffic Management
FAA	Federal Aviation Administration
FDLE	Florida Department of Law Enforcement
FDOT	Florida Department of Transportation
FTA	Florida Turnpike Authority
GOAA	Greater Orlando Aviation Authority
HNTB	Howard, Needles, Tammen, and Bergendorff
HOT	High Occupancy Toll Lanes
JPA	Joint Partnership Agreement
MPO	Metropolitan Planning Organization
M/WBE	Minority and Women's Business Enterprise
NASA	National Aeronautical and Space Agency
OIA	Orlando International Airport
OOCEA	Orlando-Orange County Expressway Authority
OSOTA	Orange, Seminole, Osceola Transportation Authority
OUATS	Orlando Urban Area Transportation Study
OTO	Office of Toll Operations
PBS&J	Post, Buckley, Schuh and Jernigan
PD&E	Project Development and Environmental
RFP	Request for Proposal
SAIC	Scientific Applications International Corporation
SCEA	Seminole County Expressway Authority
SRB	State Road Board
SRD	State Road Department
TOC	Toll Operations Contractor
VES	Vehicle Enforcement System
VTN	Voorhies, Trendle-Nelson
ZHA	Zipperly, Hardage and Associates

Bibliography

Interviews

Gerald Brinton, former Executive Director, Seminole County Expressway Authority

Gregory V. Dailer, OOCEA Chief Financial Officer, 1989-1999

Donald L. Erwin, Jr., PBS&J engineer

Jorge Figueredo, OOCEA Director of Operations, Communication and Marketing

Richard L. Fletcher, OOCEA Chairman, 1965-1971

William B. McKelvy, OOCEA Director of Construction and Maintenance

Steve Pustelnyk, OOCEA Manager of Communication and Marketing

A. Wayne Rich, OOCEA Chairman, 1992-1999

Susan Simon, OOCEA employee since 1978

Charles C. Sylvester, Jr., OOCEA Consultant 1985-1998

Harold W. Worrall, OOCEA Executive Director since 1992

Documents

Correspondence of Governor Reuben Askew, Florida State Archives, Series 136, Carton 8

Florida Statutes

Laws of Florida

OOCEA, Annual Reports, 1997-1999

OOCEA, Auditor's Report, 1967

OOCEA, Contract Files

OOCEA, Documentation Summary Manuals

OOCEA, Files of the Director of Finance

OOCEA, Financial Reports, 1985-1988

OOCEA, Five Year Work Plan, FY 2000-FY 2004

OOCEA, Long Range Expressway Plan, 1983

OOCEA, Minutes, 1967-1999

OOCEA, Response to House of Representatives, Survey of Expressway and Bridge Authorities, September 1991

OOCEA, Right-of-Way Main Parcel Acquisition Screen

OOCEA, Right-of-Way Maps

OOCEA, Right-of-Way Parcel Status Printouts, September 10, 1987

OOCEA, Western Beltway Files

Bibliography

Books, Periodicals, Pamphlets, and Scrapbooks

Asphalt Contractors Association of Florida, Newsletter, 1972

Bacon, Eve, Orlando: A Centennial History, Volume II (Chuluota, 1977)

Business Outlook, 1999

Central Florida Business, 1997

Dailer, Gregory, Donald L. Erwin, Jr., and Robert C. Hawkins, ETTM, Paper presented to IBTTA 55th Annual Meeting September 1991

International Bridge, Tunnel, and Turnpike Association, Tollways, 1989

Fletcher, Richard L., Scrapbook

Lawther, Wendell C., Privatization of Toll Operations (Expressway Authority, 1997)

OOCEA, Expressway Navigator, 1998

Orlando Business Journal, 1988-1995

PBS&J, Highlights, Summer, 1988

Powers, Ormund, Martin Andersen, Editor, Publisher, Gally Boy (Orlando, 1995)

Newspapers

Apopka Chief, 1987-1999

Atlanta Journal and Constitution, 1990

Cocoa Tribune, 1967

The Dixie Contractor, 1972-1973

Florida Today, 1987

Miami Herald, 1989

Orlando Corner Cupboard, 1965-1973

Orlando Evening Star, 1965

Orlando Sentinel, (title varies) 1965-2000

Sanford Herald, (title varies), 1974-1999

South Orlando Sun, 1990

Tampa Tribune, 1967

Winter Garden West Orange Times, 1987-1999

Footnotes

Chapter 1

1. <u>Laws of Florida</u>, Chapter 65-573; Ormund Powers, <u>Martin Andersen: Editor, Publisher, Gally Boy</u> (Orlando, 1996), pp. 250-51.

2. <u>Florida Statutes, 1995</u>, Chapter 348.754.

3. <u>Ibid.</u>, Chapter 348.7454(4).

Chapter 2

4. Orlando-Orange County Expressway Authority (OOCEA), Minutes, July 19, 1967. Additional loans of $2,500 each were made by Orlando and Orange County. The entire amount was reimbursed in 1967. OOCEA Auditor's Report, June 30, 1967, p. 5.

5. Interview with Richard L. Fletcher, July 3, 1999.

6. <u>Ibid</u>.

7. OOCEA, Minutes, July 19, through August 23, 1967.

8. Titusville <u>Star Advocate</u>, November 9, 1962, April 26, 1964, April 27, 1965; Orlando <u>Corner Cupboard</u>, July 15, 1965.

9. OOCEA, Minutes, January 15, 1964.

10. OOCEA, Minutes, January 15, 1964.

11. OOCEA, Minutes, January 15, 1964.

12. <u>Orlando Sentinel</u>, August 14, 1965. The title varies. Published as the <u>Sentinel</u> until late 1972, the newspaper was then combined with the <u>Evening Star</u>. It was published as the <u>Sentinel Star</u> until 1982 when it once again became the <u>Sentinel</u>. It is cited throughout this work as the <u>Sentinel</u>.

13. OOCEA, Minutes, June 30, 1964.

14. OOCEA, Minutes, June 30, 1964.

15. <u>Orlando Sentinel</u>, June 5, August 13, 1965.

16. Cooper was referring to the time required to pay off the bonds.

17. OOCEA, Minutes, April 21, November 11, 1965.

18. OOCEA, Minutes, June 30, September 2, December 30, 1964, April 9, 1965, February 8, 1966; <u>Orlando Sentinel</u>, January 5, 1966.

19. OOCEA, Minutes, December 30, 1964.

20. OOCEA, Minutes, December 20, 1965, May 11, 1966; <u>Orlando Sentinel</u>, October 21, 1965.

21. <u>Orlando Sentinel</u>, January 5, 1966.

22. Eve Bacon, <u>Orlando: A Centennial History</u>, Volume II (Chuluota, Mickler House, 1977), p. 270, 289.

23. <u>Orlando Sentinel</u>, July 13, 14, 1967; Ormund Powers, Martin Andersen, p. 252.

24. OOCEA, Minutes, June 29, 1967.

25. <u>Orlando Sentinel</u>, July 9, 1967.

Chapter 3

26. Quoted in <u>Orlando Sentinel</u>, April 2, 1966.

27. OOCEA, Minutes, April 1, 1966; <u>Orlando Sentinel</u>, April 1, 1966.

28. <u>Orlando Evening Star</u>, February 17, 1967.

29. OOCEA, Minutes, February 16, 1967.

30. <u>Orlando Sentinel</u>, February 17, 1967.

31. <u>Orlando Sentinel</u>, April 30, 1968.

32. <u>Orlando Sentinel</u>, July 14, 1968.

33. OOCEA, Minutes, August 27, 1968; <u>Orlando Sentinel</u>, July 14, 1968.

Footnotes

34. Martin Andersen to Richard L. Fletcher, March 8, 1973, in Richard L. Fletcher Scrapbook.

35. The fifth position on the OOCEA board was subsequently filled by an additional gubernatorial appointee until 1988 when new legislation required that it once again be filled by the Central Florida district director of the FDOT.

36. OOCEA, Minutes, June 6, 1969; Orlando Sentinel, January 23, 1969.

37. Orlando Sentinel, June 28, 1969; Orlando Evening Star, June 29, 1969.

38. Orlando Sentinel, August 11, 1969, June 2, 1974.

39. Orlando Sentinel, February 25, 1970. Governor Kirk had appointed Michael O'Neill as Transportation Secretary and the senate failed to confirm him. Cashin filled in until Ed Mueller assumed the position.

40. Orlando Sentinel, May 21, October 7, 1970; Orlando Corner Cupboard, October 10, 1970.

41. Orlando Sentinel, February 28, 1971.

42. Orlando Sentinel, May 12, 1971. This was the first bond issue to be sold with the newly authorized pledge of the full faith and credit of the state.

43. OOCEA, Minutes, November 5, 1970.

44. Orlando Sentinel, December 31, 1972.

45. OOCEA, Minutes, March 3, 1971.

46. George E. Saunders to William Owens, Office of the Governor, January 28, 1971, Florida State Archives, Series 136, Carton 8.

47. OOCEA, Minutes, February 15, 1971; Orlando Sentinel, February 16, 22, 28, 1971.

48. Orlando Sentinel, March 2, 3, 4, 1971; OOCEA, Minutes, March 4, 1971.

49. OOCEA, Minutes, March 31, 1971; Orlando Sentinel, April 23, 1971.

50. Orlando Sentinel, May 19, 1970.

51. Orlando Sentinel, April 13, 15, 16, 1971.

52. Orlando Sentinel, December 5, 10, 1970.

53. Orlando Sentinel, March 11, May 20, 1971.

54. OOCEA, Minutes, May 13, 1974; Orlando Sentinel, June 2, 1974.

55. The other agents were William E. Folmar, Emery Warner Dillard, Lewis C. Jacobs, Wardell Sims, Phillip D. Steinmetz, James A. Trutter and Joseph Underhill. Orlando Sentinel, May 16, 1971.

56. OOCEA, Minutes, April 22, September 21, October 18, 1971; Orlando Sentinel, May 21, July 27, August 18, 1971.

57. OOCEA, January 5, November 11, 1972; Orlando Sentinel, March 7, 1972.

58. Orlando Sentinel, July 15, December 3, 1971.

59. OOCEA, Minutes, August 14, 1972.

60. Orlando Sentinel, July 23, 1972.

61. Orlando Sentinel, February 16, 1973.

62. OOCEA, Minutes, October 9, November 20, 1972.

63. OOCEA, Minutes, December 21, 1971.

64. OOCEA, Minutes, January 5, 1972; Orlando Sentinel, February 6, 1972.

65. OOCEA, Minutes, October 4, 1971, April 10, May 5, August 14, 1972; Orlando Sentinel, April 11, 1971.

Footnotes

66 Orlando Sentinel, November 3, December 16, 18, 19, 20, 1972, February 12, 1973.

67 The Dixie Contractor, September 1, 1972; Orlando Sentinel, February 9, 1973.

68 Dixie Contractor, September 1, 1972.

69 Dixie Contractor, April 27, 1973.

70 Orlando Sentinel, January 9, 1973; OOCEA, Minutes, January 12, February 12, 1973.

71 OOCEA, Minutes, February 12, 1973; Orlando Corner Cupboard, February 19, 1973.

72 Orlando Sentinel, February 13, 1973.

73 Orlando Sentinel, February 16, 1973.

74 OOCEA, Minutes, February 26, 1973; Orlando Sentinel, February 21, 1973.

75 OOCEA, Minutes, February 26, 1973, Orlando Sentinel, February 27, March 16, 1973.

76 Orlando Sentinel, February 15, 1973.

77 Orlando Sentinel, January 13, 1973, February 21, 1973; OOCEA, February 26, 1973.

78 Orlando Sentinel, February 15, 1973.

79 Willard Peebles to Richard Fletcher, January 31, 1973, Martin Andersen to Fletcher, January 26, 1973, in Richard Fletcher Scrapbook.

80 Interview with Richard Fletcher, July 3, 1999.

81 Orlando Sentinel, May 20, 1971, October 5, 12, 1972; OOCEA, Minutes, September 3, 1976.

82 OOCEA, Minutes, September 16, 1974.

83 Orlando Sentinel, October 26, 27, December 11, 1973; OOCEA, Minutes, April 8, 1974.

84 OOCEA, Minutes, April 7, 1978, January 24, 1979.

85 Orlando Sentinel, July 19, 23, 1976.

Chapter 4

86 Orlando Sentinel, November 9, 1965.

87 Orlando Sentinel, November 9, 11, 1965.

88 OOCEA, Minutes, June 15, 1966; Orlando Evening Star, April 1, 1966; Orlando Sentinel, June 16, 1966.

89 Orlando Sentinel, March 16, 1967.

90 OOCEA, Minutes, June 29, 1967.

91 Laws of Florida, 1967, Chapter 65-359; Cocoa Tribune, July 7, 1967; Tampa Tribune, June 29, 1967.

92 Orlando Sentinel, January 3, 1968.

93 Orlando Sentinel, December 23, 1968.

94 Orlando Sentinel, December 20, 1968.

95 Orlando Sentinel, November 20, 1969.

96 Orlando Sentinel, May 2, 1969.

97 Orlando Sentinel, May 28, 1971.

98 Orlando Sentinel, July 11, 1971.

99 Orlando Sentinel, December 6, 1971.

100 Orlando Sentinel, March 18, August 1, December 6, 1973.

101 Orlando Sentinel, August 1, 1973.

102 Orlando Sentinel, December 14, 1973, February 17, 1974.

103 OOCEA, Minutes, August 7, 1977.

104 OOCEA, Minutes, January 4, 1977.

Footnotes

105 OOCEA, Minutes, March 19, 1980.

106 Orlando Sentinel, October 15, November 2, 1978, January 7, 1979.

107 Orlando Sentinel, August 16, November 22, 1979.

108 Orlando Sentinel, June 20, 1979; OOCEA, Minutes, April 18, 1979.

109 Interview with C. C. Sylvester, Jr., May 17, 1999; OOCEA, Minutes, January 16, February 13, 1980; Orlando Sentinel, January 22, 1981.

110 OOCEA, Minutes, March 19, 1980; Orlando Sentinel, June 20, October 10, 1979.

111 OOCEA, Minutes, December 12, 1979, January 21, 1981; Orlando Sentinel, January 22, 1981.

112 OOCEA, Minutes, January 19, 1983.

113 Orlando Sentinel, July 2, 7, 22, 1983.

Chapter 5

114 OOCEA, Minutes, November 22, December 6, 1971; Orlando Sentinel, December 7, 9, 1971.

115 Orlando Sentinel, April 20, 1972.

116 Orlando Sentinel, April 1, 20, June 1, 1972.

117 Orlando Sentinel, July 26, 1972; OOCEA, Minutes, July 1, October 9, 24, 1972.

118 Orlando Sentinel, November 2, December 5, 1972, OOCEA, Minutes, November 20, 1972, January 10, 1973.

119 Orlando Sentinel, February 25, 1973.

120 William R. Poorbaugh to Governor Askew, February 21, Askew to Poorbaugh, February 26, and Askew to J. Thomas Gurney, March 16, 1973, Florida State Archives, Series 136, Carton 8.

121 Orlando Sentinel, April 2, 1973.

122 Orlando Sentinel, September 1, November 7, 1973, January 25, 27, 1974; Sanford Herald, July 24, 1974; OOCEA, Minutes, November 1, 1974.

123 Orlando Sentinel, September 1, 1974, February 1, 1975; OOCEA, Minutes, February 1, 1975.

124 Orlando Sentinel, February 1, May 22, 1975.

125 OOCEA, Long Range Expressway Plan, 1983, p. 35, passim.

126 OOCEA, Minutes, April 18, 1979.

127 OOCEA, Minutes, May 19, 1982.

128 PBS&J Highlights, Summer 1988.

129 OOCEA, Long Range Expressway Plan, pp. 126, 127.

130 OOCEA, Long Range Plan, 1983, p. 127.

131 OOCEA, Minutes, October 12, 1983.

132 OOCEA, Minutes, May 15, December 18, 1985; Interview with Greg Dailer, July 13, 1999; Long Range Expressway Plan, June 1991, p. 2-17.

133 Orlando Business Journal, April 9, 1986.

134 Long Range Expressway Plan, June 1991, p. 2-17.

135 Orlando Sentinel, December 4, 1986; OOCEA, Minutes, April 22, 1987.

136 Orlando Sentinel, June 25, December 2, 31, 1986, January 7, 1987; Florida Today, January 9, 1987; Miami Herald, March 6, 1989; OOCEA, Financial Report, 1985, 1987, 1988.

137 Orlando Sentinel, February 9, 1988.

138 Orlando Sentinel, June 20, 1985, July 5, 1987, February 9, 1988.

Footnotes

139 Long Range Expressway Plan, June 1991, p. 2-7.

140 Orlando Sentinel, January 4, 1985.

141 Orlando Sentinel, January 18, 1985.

142 Orlando Sentinel, March 25, October 7, 1987.

143 International Bridge, Tunnel and Turnpike Association, Inc., Tollways, June 1989.

144 OOCEA, Minutes, November 21, 1984, April 17, 1985; Orlando Sentinel, February 17, 1988.

145 OOCEA, Minutes, January 20, 1985; Orlando Sentinel, February 9, 1985

146 OOCEA, Minutes, January 20, 1985; April 27, 1988.

Chapter 6

147 OOCEA, Long Range Expressway Plan, 1983, pp. 113-116.

148 OOCEA, Minutes, January 25, 1984.

149 OOCEA, Minutes, June 25, 1986; Orlando Sentinel, April 11, 1987.

150 OOCEA, Minutes, February 22, 1984.

151 Orlando Sentinel, February 22, 1984.

152 Orlando Sentinel, April 10, 13, 17, 26, 1984

153 Orlando Sentinel, May 28, June 1, September 30, 1984.

154 Orlando Sentinel, June 24, 1984.

155 Orlando Sentinel, June 24, 1984, August 21, 1985; OOCEA, Minutes, August 7, 1985.

156 OOCEA, Minutes, April 17, 1985.

157 Orlando Sentinel, September 20, 1986, and November 14, 1985, February 26, 1986; OOCEA, Minutes, December 18, 1985.

158 OOCEA, Minutes May 27, 1987.

159 OOCEA, Minutes, May 24, 1989.

160 OOCEA, Minutes, June 19, August 28, 1985.

161 OOCEA, Minutes, December 18, 1985, January 21, August 20, 1986.

162 Interview with Charles C. Sylvester, Jr., July 29, 1999.

163 Orlando Sentinel, May 10, 1988; Northeastern Beltway Right-of-way Maps; OOCEA Parcel Status Printouts, September 10, 1987.

164 Orlando Business Journal, September 4-10, 1988.

165 Interview with William McKelvy, July 21, 1999.

166 OOCEA, Minutes, April 22, 1987, October 26, 1988.

167 OOCEA, Minutes, January 21, 1987.

168 OOCEA, Minutes, May 27, 1987.

169 OOCEA, April 22, 1987.

170 Orlando Sentinel, July 15, 1987.

171 Project 101, Documentation Summary Manual, OOCEA files; OOCEA, Minutes, May 27, August 27, 1987.

172 Project 102, Documentation Summary Manual, OOCEA files; OOCEA, Minutes, July 29, August 26, 1987.

173 Project 103, Documentation Summary Manual, OOCEA files; OOCEA, Minutes, June 17, July 29, August 26, 1987.

174 Project 104, Documentation Summary Manual, OOCEA files; OOCEA, Minutes, August 26, 1987.

Footnotes

175 OOCEA, Minutes, July 29, September 30, 1987; Project 105, Documentation Summary Manual, OOCEA, files.

176 Orlando Sentinel, June 22, September 19, 1986.

177 Orlando Sentinel, December 27, 1988; Interview with Gerald Brinton, September 29, 1999.

178 Project 106, Documentation Summary Manual, OOCEA files; OOCEA, Minutes, October 28, 1987.

179 Orlando Sentinel, October 20, 1987, October 2, 1988.

180 OOCEA, Minutes, November 23, December 22, 1988, April 26, 1989; Orlando Sentinel, December 16, 1988; OOCEA, Response to Florida House of Representatives, Survey of Expressways and Bridge Authorities, September 1991.

Chapter 7

181 OOCEA, Long Range Expressway Plan, 1983, p. 127.

182 Orlando Sentinel, April 16, 1982; Orlando, the little Sentinel (a section of the Orlando Sentinel), February 25, 1983.

183 OOCEA, Minutes, July 25, 1984.

184 OOCEA, Minutes, September 19, November 21, 1984.

185 OOCEA, Minutes, August 26, 1987.

186 OOCEA, Right-of-Way Maps, Projects 304/305 and Parcel 40-110 Folder; Interview with Charles C. Sylvester, July 26, 1999.

187 OOCEA, Minutes, June 22, 1986; Orlando Sentinel, July 15, 20, August 25, September 16, 1986.

188 OOCEA, Right-of-Way Maps, Projects 301-305. Orlando Sentinel, February 13, 15, 1987; Orlando Business Journal, July 24-30, 1988.

189 Orlando Sentinel, March 3, 1989.

190 OOCEA, Minutes, December 18, 1985, January 21, March 13, 1986;

191 OOCEA, Minutes, July 23, September 30, 1987.

192 OOCEA, Minutes, September 30, 1987; Project 301, Documentation Summary Manual, July, 1988.

193 OOCEA, Minutes, October 28, 1987; Project 302. Documentation Summary Manual.

194 OOCEA, Minutes, December 23, 1987; Project 3303, Documentation Summary Manual, OOCEA Files.

195 OOCEA, Minutes, November 25, 1987; Project 304, Documentation Summary Manual.

196 OOCEA, Minutes, December 23, 1987.

197 Project 306, Documentation Summary Manual.

198 OOCEA, Minutes, April 27, 1988.

199 Orlando Sentinel, May 13, June 12, 1989.

Chapter 8

200 OOCEA, Long Range Expressway Plan, 1983, p. 127.

201 OOCEA, Minutes, July 25, 1984; Orlando Sentinel, April 23, 1987.

202 OOCEA, Minutes, July 25, 1984; Orlando Sentinel, November 3, 1986.

203 OOCEA, Minutes, February 20, December 12, 1985, January 21, 1986.

Footnotes

[204] OOCEA, Minutes, July 23, 1986; <u>Orlando Sentinel</u>, April 23, 1987. After the environmental problems arising from the Northeastern Beltway project, at the suggestion of Phil Reece and his fellow board members, Bill Gwynn had employed Hal Scott, a former Audubon Society president, as a consultant to work with a nine member panel of conservationists to advise on route selections. <u>Orlando Sentinel</u>, February 7, 1988.

[205] OOCEA, Minutes, November 12, 1986, January 21, 1987.

[206] OOCEA, Minutes, March 23, June 22, 1988.

[207] OOCEA, Parcel Files for Projects 40l/402A, 404, and 405.

[208] <u>Orlando Sentinel</u>, March 21, 1988.

[209] Interview with Charles C. Sylvester, July 26, 1999.

[210] OOCEA, Minutes, August 24, September 27, December 22, 1988; Parcels 4l-101, 41-102, 41-105, 41-106, Project 405 Files; Interview with Charles C. Sylvester, July 26. 1999.

[211] OOCEA, Minutes, April 27, August 24, 1988, May 24, 1989.

[212] OOCEA, Minutes, November 23, 1988.

[213] OOCEA, Minutes, December 22, 1988; Projects 40l/402A, Documentation Summary Manual, OOCEA Files.

[214] OOCEA, Minutes, March 22, 1989.

[215] OOCEA, Minutes, October 11, 1988.

[216] <u>Orlando Sentinel</u>, April 17, 1990.

[217] <u>Orlando Sentinel</u>, June 7, 1990.

[218] <u>Orlando Sentinel</u>, June 23, 1990; OOCEA, <u>Response to Florida House of Representatives, Committee on Transportation, Survey of Expressway and Bridge Authorities</u>, September 1991.

Chapter 9

[219] <u>Orlando Sentinel</u>, March 21, 1985.

[220] OOCEA, August 28, 1985; <u>Orlando Sentinel</u>, June 26, 1985.

[221] <u>Orlando Sentinel</u>, August 8, 1985.

[222] OOCEA, Minutes, September 17, 1985.

[223] <u>Orlando Sentinel</u>, January 22, 1986; OOCEA, Minutes, January 21, 1986.

[224] <u>Orlando Sentinel</u>, April 4, March 14, 1986; OOCEA, Minutes, March 13, 1986.

[225] <u>Orlando Sentinel</u>, April 4, March 14, 1986; OOCEA, Minutes, March 13, April 17, 1986.

[226] OOCEA, Minutes, July 23, 1986.

[227] OOCEA, Minutes, August 20, 1986; <u>Orlando Sentinel</u>, August 20, 21, 1986.

[228] <u>Orlando Sentinel</u>, October 3, 1986.

[229] OOCEA, Minutes, November 12, 1986, February 18, 1987.

[230] OOCEA, Minutes, June 17, August 26, 1987, May 25, March 22, 1989.

[231] OOCEA, Parcel Status Printout, November 3, 1988.

[232] OOCEA, Minutes, October 11, November 23, December 22, 1988.

[233] OOCEA, Minutes, April 27, August 24, 1988.

Footnotes

234 OOCEA, Minutes, April 26, 1989; Project 502, Documentation Summary Manual, OOCEA Files.

235 OOCEA, Minutes, February 22, 1989; Projects 504/505, Documentation Summary Manual, OOCEA Files.

236 OOCEA, Minutes, May 24, 1989; Project 501, Documentation Summary Manual, OOCEA Files.

237 OOCEA, Minutes, July 24, 1989.

238 OOCEA, Response to Florida House of Representatives, Committee on Transportation, Survey of Expressway and Bridge Authorities, September 1991.

239 OOCEA, Minutes, March 27, May 22, 1991.

Chapter 10

240 OOCEA, Minutes, October 11, 1988. Recognizing that the road needs for the Orlando area far exceeded anticipated revenues available for them, the MPO had reduced its listing of roads for 2005 to only those which it deemed financially feasible. The Central Connector was important enough to be included on that selective list.

241 Orlando Sentinel, April 28, 1988.

242 OOCEA, Minutes, August 24, 1988.

243 OOCEA, Minutes, August 24, 1988; Orlando Sentinel, August 25, 1988.

244 Orlando Business Journal, December 17-23, 1989.

245 Quoted in Orlando Sentinel, September 10, 1988.

246 Orlando Sentinel, October 11, 27, 1988.

247 OOCEA, Minutes, October 11, 1988. Barton-Aschman Associates, Inc., had just completed a contract to make the study in association with DeLeuw, Cather & Company and Glatting, Lopez, Kercher, Anglin, Inc., for $1,200,000. Their report was due in October 1989.

248 As quoted by the Orlando Sentinel, October 12, 1988.

249 Orlando Sentinel, October 12, 1988.

250 Orlando Sentinel, February 25, 1989; Orlando Business Journal, July 23-29, 1989.

251 Florida Statutes, 348.754(4)

252 This is exactly as quoted in the Orlando Sentinel, January 18, 1989.

253 Orlando Sentinel, February 25, April 13, 1989.

254 Orlando Sentinel, April 9, 1989.

255 OOCEA, Minutes, April 26, May 24, June 28, 1989; Orlando Business Journal, June 18-24, 1989; Orlando Sentinel, August 10, 17, 1989.

256 Orlando Sentinel, April 24, 1989.

257 As quoted in Orlando Sentinel, August 14, 1989.

258 OOCEA, Minutes, August 23, 1989; Orlando Sentinel, August 24, September 12, 1989.

259 Orlando Sentinel, September 13, 1989.

260 Orlando Sentinel, October 5, 1989; OOCEA, Minutes, October 25, November 22, 1989.

261 OOCEA, Minutes, April 24, June 26, 1991.

262 Orlando Sentinel, October 26, November 23, December 29, 1989; Orlando Business Journal, November 19-25, 1989.

Footnotes

263 Orlando Sentinel, January 21, June 15, 1990.

264 Orlando Sentinel, October 3, 1989.

265 OOCEA, Minutes, August 22, 1990, January 23, March 27, April 24, June 26, July 24, 1991, May 27, 1992.

266 Orlando Sentinel, April 24, 26, May 2, 30, 1991.

267 The South Orlando Sun, September 6, 1990; Orlando Sentinel, October 16, 1990, August 16, 1991.

268 OOCEA, Minutes, November 13, 1991.

269 OOCEA, Minutes, November 20, 1991.

270 OOCEA Central Connector, 1988 Project Costs, Files of OOCEA Director of Finance; Orlando Sentinel, March 26, 1992; Orlando Business Journal, April 9-15, 1993.

271 Orlando Sentinel, November 14, 1991, January 26, February 29, March 4, 14, December 15, 1992, January 22, March 30, 1993.

271A Surface Transportation Alternatives for the Orlando Urban Area South-Central Corridor Mediation, July, 1995.

Chapter 11

272 Orlando Sentinel, January 21, 1990.

273 Orlando Sentinel, January 21, 1990.

274 Orlando Sentinel, January 24, 1990.

275 Orlando Sentinel, February 23, 1990.

276 Orlando Sentinel, February 23, 1990; Bill Gwynn to Authority Members, October 23, 1990, Exhibit B in OOCEA, Minutes, October 23, 1990.

277 Orlando Sentinel, February 4, 1990.

278 Orlando Sentinel, March 22, 23, 1990.

279 As quoted in Orlando Sentinel, October 14, 1990.

280 Orlando Sentinel, July 10, October 2, 3, 14, 1990.

281 Orlando Sentinel, October 16, 17, December 1, 1990, October 15, 1991.

282 Orlando Sentinel, December 7, 1990.

283 As quoted in Orlando Sentinel, January 24, 1991.

284 OOCEA, Minutes, January 23, 1991; Orlando Sentinel, February 5, 1991.

285 OOCEA, Minutes, February 27, 1991; Orlando Business Journal, May 31-June 6, 1991.

286 OOCEA, Minutes, April 4, September 25, 1991; Orlando Sentinel, April 23, 1991.

287 As quoted in Orlando Sentinel, May 27, 1991.

288 Orlando Sentinel, May 27, 1991.

289 Orlando Sentinel, June 13, 1991.

290 OOCEA, Minutes, October 9, 1991.

291 OOCEA, Minutes, August 28, 1991, February 2, May 27, 1992.

292 Orlando Sentinel, October 10, 1991.

293 OOCEA, Minutes, October 23, 1991.

294 OOCEA, Minutes, March 25, 1992; Orlando Sentinel, September 26, 1991.

295 Orlando Sentinel, May 14, 1991.

296 Orlando Sentinel, July 15, August 9, 1991, OOCEA, Minutes, June 26, August 8, December 18, 1991, March 25, 1992.

297 OOCEA, Minutes, November 28, 1990; February 24, March 24, May 26, 1993.

Footnotes

298 South Orlando Sun, January 11, March 15, May 3, 1990; OOCEA, Minutes, February 28, 1990; Orlando Sentinel, March 16, June 18, 1990. Durek's suit was subsequently dismissed.

299 Orlando Sentinel, September 24, 26, December 27, 1990; West Orange Times, February 14, 1991.

300 As quoted in Orlando Sentinel, February 28, 1991. See also Orlando Sentinel, July 22, 1992.

301 OOCEA, Minutes, January 22, 1992.

302 OOCEA, Minutes, July 22, 1992; Orlando Sentinel, July 23, 1992.

303 OOCEA, Minutes, July 22, 1992, March 24, 1993; Orlando Sentinel, August 27, 1992.

304 OOCEA, Minutes, February 24, April 28, June 17, 1993; Orlando Sentinel, May 27, 1993.

305 OOCEA, Minutes, June 7, September 12, 1990.

306 Orlando Sentinel, March 2, 1992.

307 OOCEA, Minutes, December 18, 1991, January 22, February 26, 1992, April 1, 1993.

308 OOCEA, Minutes, June 23, November 17, 1993.

309 Gregory V. Dailer to Ranson Smith, June 26, 1997, in Files of the OOCEA Director of Finance; Orlando Sentinel, June 10, 1993.

310 OOCEA, Minutes, February 27, 1991.

311 OOCEA, Minutes, March 27, 1991.

312 Orlando Sentinel, March 28, April 4, 1991; OOCEA, Minutes, May 22, 1991.

313 Orlando Sentinel, June 28, 1992.

314 OOCEA, Minutes, April 28, 1993.

315 Orlando Sentinel, December 18, 1991, January 23, February 19, 1992; OOCEA, Minutes, August, 26, December 23, 1992.

316 Asphalt Contractors Association of Florida, Newsletter, April 1992; OOCEA, Minutes, August 26, 1992.

317 OOCEA, Minutes, August 26, 1992.

Chapter 12

318 Nelson R. Boice to James B. Greene, March 15, 1974, in possession of Charles C. Sylvester, Jr.

319 The four owners were: Sunley Holding of America (Lake Nona, with 6750 acres), Sonny Lawson of Kissimmee with 1,500 acres, Landstar Homes (Meadow Woods, with 3,000 acres), and Richland Properties of Florida, Inc., (Southchase, with 3,200 acres).

320 Orlando Sentinel, March 16, 1986.

321 OOCEA, Minutes, September 19, 1984.

322 Orlando Sentinel, March 9, 1984, August 24, 1987.

323 Orlando Sentinel, June 15, 1983.

324 Orlando Sentinel, September 2, 1986.

325 OOCEA, Minutes, March 18, 1987; Orlando Sentinel, April 13, 1987.

326 Orlando Sentinel, October 13, 1987.

327 Orlando Sentinel, November 1, December 10, 11, 24, 1987, January 11, 13, 14, 18, 19, 20, 23, February 25, 1988; OOCEA, Minutes. November 25, December 23, 1987, January 27, April 27, May 25, 1988.

328 OOCEA, Minutes, June 22, 1988.

329 Orlando Sentinel, June 20, 1988.

Footnotes

330 Orlando Sentinel, July 1, 1989.

331 Orlando Sentinel, March 30, 1989.

332 Orlando Sentinel, October 18, 1989; OOCEA, Minutes, October 25, 1989.

333 OOCEA, Minutes, March 28, 1990.

334 OOCEA, Minutes, March 28, 1990.

335 OOCEA, Minutes, November 22, 1989.

336 OOCEA, Minutes, March 28, December 19, 1990, March 27, 1991.

337 Orlando Sentinel, December 4, 1988. May 11, November 11, 1990; OOCEA, Minutes, September 25, 1991. Other projects included in the Beltway Mitigation Bill included the Western Beltway, the Southern Connector Extension, and the Turnpike.

338 OOCEA, Minutes, November 20, 1992, February 24, 1993.

339 As quoted in Orlando Sentinel, September 25, 1990.

340 OOCEA, Minutes, July 25, 1990; Orlando Sentinel, July 26, 31, August 23, September 25, 1990; South Orlando Sun, July 19, 1990.

341 OOCEA, Minutes, November 13, 1990; Orlando Sentinel, December 7, 1990.

342 Interview with Charles C. Sylvester, Jr., July 26, 1999.

343 Interview with Charles C. Sylvester, July 26. 1999; Printout of Parcel Status Report, September 6, 1991; Main Parcel Acquisition Screen, September 21, 1999; Orlando Sentinel, October 16, 20, 1992.

344 OOCEA, Minutes, April 24, June 12, 1991.

345 OOCEA, Minutes, August 8, 1991.

346 OOCEA, Minutes, August 28, September 25, 1991, February 26, 1992.

347 OOCEA, Minutes, September 25, 1991; Project 450, Documentation Summary Manual.

348 OOCEA, Minutes, August 28, 1991; Project 451, Documentation Summary Manual.

349 OOCEA, Minutes, September 1991; Project 453, Documentation Summary Manual.

350 OOCEA, Minutes, October 23, 1991; Project 454, Documentation Summary Manual.

351 OOCEA, Minutes, August 28, 1991; Project 455, Documentation Summary Manual.

352 OOCEA, Minutes, July 24, 1991; Project 457, Documentation Summary Manual.

353 Project 456, Documentation Summary Manual.

354 OOCEA, Minutes, November 20, 1991, February 26, 1992.

355 Orlando Sentinel, February 7, 1993.

356 OOCEA, Minutes, February 23, 1994.

357 OOCEA, Minutes, June 7, 1995.

358 OOCEA, Minutes, June 23, 1994; Orlando Sentinel, January 10, 1993.

359 OOCEA, Minutes, April 28, 1993; Orlando Sentinel, August 1, 1992.

360 OOCEA, Minutes, March 24, May 26, June 23, 1993; Orlando Sentinel, June 17, 24, 1993.

361 Orlando Sentinel, July 25, 1993.

362 Orlando Sentinel, July 24, 1992.

363 Orlando Sentinel, August 18, 1995, and Central Florida Business, May 5-11, 1997; OOCEA, Minutes, August 25, 1999.

364 Orlando Sentinel, July 24, 25, 1992, December 29, 1993. The Disney World officials were not just being generous. The company had plans for new theme parks and hotels as well as the new city of Celebration. It was unlikely that they could have obtained state approval for those projects without considerable improvement in area roads.

365 Orlando Sentinel, June 17, 1996, June 23, 1990.

366 Orlando Sentinel, November 3, 1989; April 7, May 25, June 23, October 11, 1990; Interview with Gerald Brinton, September 29, 1999.

367 Orlando Sentinel, September 9, 1993.

368 Orlando Sentinel, April 26, 1994.

369 Orlando Sentinel, September 9, 1993.

370 Orlando Sentinel, September 21, 1993, January 12, April 1, 5, May 18, 1994; Orlando Business Journal, December 31, January 6, 1994.

371 Orlando Sentinel, May 12, September 23, 1994, July 10, August 21, November 24, 1995, March 22, 1996, August 13, 1999, October 12, 1999; Interview with Gerald Brinton, September 29, 1999.

Chapter 13

372 OOCEA, Minutes, April 27, 1988.

373 OOCEA, Minutes, March 22, 1989.

374 Atlanta Journal and Constitution, August 18, 1990.

375 Gregory V. Dailer, Donald L. Erwin, Jr., and Robert C. Hawkins, ETTM System. (Paper presented to the International Bridge, Tunnel, and Turnpike Association 59th Annual Meeting, September 29-October 3, 1991), p. 4.

376 Dailer, et al, passim.

377 Interview with Donald L. Erwin, Jr., November 5, 1999; OOCEA, Minutes, October 24, 1990.

378 OOCEA, Minutes, March 27, June 12, July 24, 1991; Interview with Donald L. Erwin, Jr., November 5, 1999.

379 OOCEA, Minutes, August 28, 1991.

380 Orlando Business Journal, November 8-14, 1991.

381 Interview with Donald L. Erwin, Jr., November 5, 1999.

382 OOCEA, Minutes, June 24, July 22, 1992; Letter of Dr. Lorenz A. Kull to Robert A. Mandell, June 3, 1992.

383 OOCEA, Minutes, July 22, December 23, 1992.

384 OOCEA, Minutes, July 22, 1992.

385 OOCEA, Minutes, October 28, 1992.

386 Orlando Sentinel, November 15, 1992, March 29, 1993.

387 Supplemental Agreement No. 4, January 15, 1993; OOCEA, Minutes, January 15, 1993; Wendell C. Lawther, Privatization of Toll Operations of Orlando Orange County Expressway Authority, (Expressway Authority, 1997), p. 17.

388 OOCEA, Minutes, April 14, May 11, 26, 1993.

Footnotes

389 Interviews with Donald L. Erwin, Jr., November 5, 1999 and Harold Worrall, November 18, 1999.

390 OOCEA, Minutes, September 28, 1994.

391 OOCEA, Annual Report, 1995, p. 24; Interview with Harold Worrall, February 14, 2000.

392 Orlando Sentinel, May 2, 1994; OOCEA, Minutes, November 18, 1992.

393 Orlando Sentinel, May 2, 1994.

394 Orlando Sentinel, May 2, 1994.

395 OOCEA, Minutes, March 22, 1995, November 28, 1996.

396 OOCEA, Minutes, May 25, 1994; Orlando Sentinel, October 24, 1994; January 16, 1995.

397 Orlando Sentinel, November 22, 1994.

398 Orlando Sentinel, April 30, May 1, 1995.

399 Florida Statutes, s. 338.155(l); OOCEA, Minutes, April 28, 1993.

400 OOCEA, Minutes, August 24, 1994; Orlando Sentinel, July 3, 1995.

401 Orlando Sentinel, April 21, 24, 1997.

402 Orlando Sentinel, May 17, 1995.

403 Orlando Sentinel, January 12, 27, August 16, September 9, 1996.

404 OOCEA, Minutes, August 23, 1995; Orlando Sentinel, December 8, 1995, January 27, 1996.

405 Orlando Sentinel, June 26, 1997; Interview with Jorge Figueredo, November 22, 1999.

406 Interview with Jorge Figueredo, November 22, 1999.

407 Interviews with Harold Worrall, November 18, 1999, Jorge Figueredo, November 22, 1999, and Steve Pustelnyk, August 2, 2000.

Chapter 14

408 Lawther, Privatization of Toll Operations, p. 9, 19.

409 OOCEA, Minutes, January 23, 1991.

410 OOCEA, Minutes, August 28, 1991.

411 OOCEA, November 20, 1991, February 26, April 22, 1992; Lawther, Privatization of Toll Operations, pp. 12-13.

412 Lawther, Privatization of Toll Collections, p.17.

413 Lawther, Privatization of Toll Collections, pp. 21-22.

414 Lawther, Privatization of Toll Operations, p. 85, passim. Unless otherwise noted, the information in the following pages is based on this work. Citations have been added only when specific page number identification seemed appropriate.

415 Lawther, Privatization of Toll Operations, p. 85.

416 Lawther, Privatization of Toll Operations, pp. 87-89.

417 For a detailed analysis of the Project 266 RFP, see Lawther, Privatization of Toll Operations, Chapter 5.

418 OOCEA, Minutes, March 23, 1994.

419 Lawther, Privatization of Toll Operations, pp. 24-25.

420 Lawther, Privatization of Toll Operations, pp. 25-26.

421 Lawther, Privatization of Toll Collections, p. 27.

422 OOCEA, Minutes, November 11, 1994, June 7, 1995; Orlando Business Journal, May 5-11, 1995.

423 OOCEA, Minutes, January 25, 1995.

Footnotes

424 Lawther, Privatization of Toll Collections, pp. 29, 247, 252.

425 OOCEA, Minutes, August 24, 1994, June 7, 1995.

426 OOCEA, Minutes, June 23, 1999.

Chapter 15

427 OOCEA, Long Range Expressway Plan, 1983, p. 129.

428 OOCEA, Minutes, July 23, 29, December 17, 1986; Orlando Sentinel, January 12, 1987; Apopka Chief, January 16, 1987; Winter Garden West Orange Times, October 8, 1987.

429 OOCEA, Western Beltway, Part A. December 5, 1995; Gary Skaff (PBS&J) to Bruce Barrett (FDOT), December 5, 1991, OOCEA Western Beltway Files.

430 Orlando Sentinel, November 3, 1987.

431 OOCEA, Minutes, June 22, August 24, 26, 1988.

432 Orlando Sentinel, February 22, 23, 1989.

433 Orlando Sentinel, May 20, 29, June 5, 1989.

434 Orlando Sentinel, June 3, 4, 5, 1989.

435 Orlando Sentinel, November 3, 12, 19, 1989.

436 Orlando Sentinel, June 9, 1990; OOCEA, Minutes, May 22, 1991.

437 Robert J. Paulsen (PBS&J) to D.W. Gwynn, June 18, 1991, OOCEA Western Beltway Files.

438 Jewel Symmes to Allan Weber, March 23, 1992, OOCEA Western Beltway Files; Orlando Sentinel, November 21, 1991.

439 Orlando Sentinel, December 12, 1991; Winter Garden West Orange Times, November 21, 1991.

440 Orlando Sentinel, November 2, 1993.

441 Orlando Sentinel, November 2, 1993.

442 Florida Statutes, 1994, Section 348.7545; FDOT, Preliminary Feasibility Report: Western Beltway, Parts A, B and C, May 1994, OOCEA Western Beltway Files.

443 Orlando Sentinel, October 18, 1992; Apopka Chief, May 13, 1994.

444 Beltway 2000 Task Force, Memorandum of Meeting, June 3, 1994, OOCEA Western Beltway Files; Apopka Chief, September 9, 1994; Orlando Sentinel, September 11, 1994.

445 A. Wayne Rich, "Creating New Solutions for Central Florida's Transportation Needs," September 1994.

446 OOCEA, 2015 Master Plan, pp. 6-2, 6-4.

447 OOCEA, 2015 Master Plan, p. 6-9.

448 OOCEA, 2015 Master Plan, p. 6-12.

449 OOCEA, 2015 Master Plan, p. 6-13.

450 OOCEA, Minutes, December 18, 1985.

451 Orlando Sentinel, January 26, 1994.

452 Winter Garden West Orange Times, August 5, 1999; Apopka Chief, July 30 1999, Sanford Seminole Herald, August 1, 1999.

453 Harold Worrall to Task Force Members, January 27, 1995, OOCEA Western Beltway Files.

454 OOCEA, Minutes, January 15, February 28, June 7, 1995.

455 OOCEA, Minutes, June 7, 1995; Western Beltway Policy Committee and Project Task Force, Agenda, August 31, 1995, OOCEA Western Beltway Files.

Footnotes

456 OOCEA, Minutes, November 22, 1995; Winter Garden West Orange Times, May 16, 1996; Apopka Chief, May 10, 1996.

457 Harold W. Worrall to John R. Freeman, Jr., September 8, 1997, OOCEA Western Beltway Files; Interview with Hal Worrall, February 3, 2000.

458 OOCEA, Minutes, January 24, August 28, 1996.

459 OOCEA, Minutes, June 26, October 30, 1996; Winter Garden West Orange Times, November 7, 1996; Orlando Sentinel, October 31, 1996.

460 OOCEA, Minutes, February 28, 1995.

461 OOCEA, Minutes, June 7, October 25, 1995.

462 OOCEA, Minutes, October 25, November 21, 1995; Apopka Chief, June 9, December 1, 1995.

463 Apopka Chief, May 10, 1996; Orlando Sentinel, March 28, 1996; December 30, 1997.

464 OOCEA, Contract Files; Winter Garden West Orange Times, January 16, 1997.

465 Florida Statutes, 1995, section 338.2276.

466 Sarah Bleakley to Honorable Daniel A. Webster, June 20, 1997, OOCEA Western Beltway Files.

467 OOCEA, Minutes, March 25, 1998.

468 OOCEA, Minutes, February 18, March 25, 1998.

469 OOCEA, Minutes, December 17, 1997, August 26, 1998.

470 OOCEA, Minutes, November 19 1997; OOCEA Contract Files.

471 OOCEA, Minutes, May 27, 1998.

472 OOCEA, Minutes, May 27, 1998; Orlando Sentinel, June 1, 1998; Business Outlook, April, 1999.

473 OOCEA, Minutes, September 23, 1998.

474 OOCEA, Minutes, October 28, 1998.

475 OOCEA, Minutes, June 23, 1999.

476 OOCEA, Minutes, June 7, 1995; Florida Statutes, 1995, section 348.760.

477 Orlando Sentinel, February 7, 8, May 24, 1995; Western Beltway Task Force Meeting, Minutes, February 14, 1995, Brian C. Canin to Tom Ross, March 31, 1995, C. David Brown to Brady Sneath, April 9, 1996, OOCEA Western Beltway Files.

478 C. David Brown to Brady Sneath, April 9, 1996, OOCEA Western Beltway Files.

479 OOCEA, Minutes, April 22, 1998.

480 OOCEA, Minutes, April 22, 1998; Orlando Sentinel, April 7, 1998.

481 OOCEA, Minutes, November 30, 1998, Exhibit B; Orlando Sentinel, April 8, 1998; Winter Garden West Orange Times, April 9, 1998; Thomas F. Barry, Jr., to A. Wayne Rich, et al, Memorandum, April 27, 1988, OOCEA Western Beltway Files.

482 Joint Partnership Agreement for the Preconstruction Activities of Western Beltway Part C, OOCEA Western Beltway Files; Business Outlook, April, 1999.

483 Interview with Harold Worrall, February 2, 2000.

484 Apopka Chief, January 24, 1997.

Footnotes

Chapter 16

485 OOCEA, Minutes, May 27, 1998.

486 OOCEA, Minutes, March 26, 1997, November 12, 1999; Orlando Sentinel, November 13, 1999; Interview with Steve Pustelnyk, August 2, 2000.

487 OOCEA, Five Year Work Plan, FY 2000-2004, Section 6.

488 OOCEA, Minutes, February 26, 1997.

489 OOCEA, Minutes, May 26, 1999.

490 OOCEA, Minutes, May 26, June 23, 1999.

491 OOCEA, 2015 Master Plan, pp. 6-2, 6-4.

492 OOCEA, Minutes, June 26, 1996, January 22, 1997; OOCEA, Annual Report, 1997, p. 4.

493 Orlando Sentinel, May 18, 1997.

494 OOCEA, Minutes, November 30, 1998, Exhibit F.

495 OOCEA, Minutes, November 30, 1998.

496 OOCEA, Minutes, February 24, 1999.

497 OOCEA, Five Year Work Plan, FY 2000 - FY 2004, p. 37.

498 OOCEA, Minutes, March 24, April 28, 1999.

499 OOCEA, April 28, 1999.

500 Interview with Jorge Figueredo, February 14, 2000.

501 Orlando Sentinel, May 29, 1997; OOCEA, Minutes, May 28, 1997; Florida Statutes, 1997, section 348.760.

502 OOCEA, Minutes, May 28, 1997; Orlando Sentinel, May 23, 1997.

503 OOCEA, Minutes, November 1999.

504 Orlando Sentinel, August 16, 1999.

505 OOCEA, Minutes, September 22, November 12, 1999.

506 Harold Worrall, "A Strategic Analysis of the Orlando-Orange County Expressway Authority," March 1996.

507 OOCEA, Minutes, March 26, 1997.

508 OOCEA, 2025 Long Range Master Plan, pp. 6-11, 6-12.

509 OOCEA, 2025 Long Range Master Plan, pp. 6-12-16-13.

Index of Key Words

1986 Project 4, 28, 67, 69, 71, 72, 90, 95, 131, 155-156, 186
2015 Master Plan . . . . .188 - 190, 200, 205, 209
2025 Long Range Master Plan .216

A

Airport Expressway 61, 64
Airport Interchange . . . . .52 - 56, 60, 66
Airport Road 151
Akerman, Senterfitt & Edison 133, 193
Alafaya Trail 87, 89, 91, 92, 109
American Newland Associates 143
Andersen, Martin 6, 7, 10, 16, 17, 24, 41, 132
Anderson Street 21, 22, 31
Andrews, Craig 119
Apopka 64, 154, 181, 182, 186, 187, 189, 190, 194, 212
Apopka Bypass 183,186, 189, 190, 192, 193
Arthur, Allen 69, 106
Askew, Reuben 27, 35, 39, 40, 34, 59
Asphalt Pavers, Inc. 108
Automated Vehicle Identification (AVI) 157-161, 163-165, 167-173, 175, 177, 213
Azalea Park 32

B

BDA Environmental Consultants 141
BMK Ranch 141
Ballenger and Company 33, 37
Banks, E.G. 80
Barley, Sr., George M. 69, 106
Barlow, Angus 29
Barnes, Raymond E. 9, 11
Barnhill, Bruce 30
Barr, Jacqueline D. 164, 212
Barry, Thomas 125, 133
Barton, Don 54
Barton-Aschman Associates 113, 115
Bates, William J. 32
Bay Run Subdivision 90

Beard, Art 15
Beckett, William A. 212, 213
Beiswenger, Hoch & Associates, Inc. 106, 140
Belle Isle 114
Beltway 2000 Task Force .186, 190
Beltway General Mitigation Bil 140
Bennett Causeway 45-47
Benton, Al 89
Berenis, Joseph A. 71, 72, 119, 123, 128, 139, 156, 160, 191, 192, 212
Bergendorff 10, 21, 140, 213
Bergeron Land Development, Inc. 145
Berry Dease Road 100
Bertossa, Harry 24, 29, 32, 39, 42, 59, 60, 213
Bethlehem Steel 37
Blozak, M.G. 17
Bob Carr Performing Arts Center 75
Bobber, Dorris 115-117, 119
Boggy Creek Road 52, 139, 142, 144-148
Boh Brothers 152, 153
Boice, Nelson 13, 135
Bond 8, 10-13, 22, 24, 26-30, 43, 46-48, 52-54, 57, 60, 66-68, 73, 90, 99, 106, 112, 130-131, 137-138, 141-142, 150-152, 183-184, 186, 196, 207
Bond Review Board 12
Bond sale 28, 29, 48, 66-68, 106, 112, 142
Bowyer-Singleton & Associates, Inc. 116, 195
Bradshaw, Jr., Charles E. 143
Brevard County .7, 10, 11, 45, 47, 49
Brevard County Commission 45, 47
Brewer, Max 9, 10, 12, 14, 17
Bridge 8, 13-14, 25-26, 29-30, 34, 37-38, 83-84, 90, 106, 144-146, 152-153, 173, 197, 198, 206
Brinton, Gerald 4, 74, 169
Broad & Cassel 133, 193, 209
Brown, David 193, 209
Brown, Jay 17, 46, 47
Brunetti, J. J. 13
Brunetti, John 96, 99

Index of Key Words

Bryan, Norman 15
Bryan Road 82
Bryant, Farris 8
Buie, Suzanne 30
Building Movers, Inc. .32, 39, 59
Bumby Avenue 32, 68
Bumper-Mounted 210
Burger, Milissa 174, 178
Burkhard, Phil 171
Burns, Haydon 18, 20, 25

C
CCTV 157, 167
CEI 81, 82, 91, 92, 100, 107-108, 116, 141, 144
CSX Transportation .113, 144, 145
Callahan, Mark 191
Campbell, Foxworth and Pugh 57
Cape Canaveral 5-8, 10-11, 17-18, 50
Cargo Road 208
Carr, Robert 16
Carter, Maury 146
Carter, Vera 141-142
Central Connector 61, 64-65, 109, 111-113, 115-120,
121, 125, 129, 131, 185
Central Florida Development Commission (CFDC) . . . . . .
. . . . .7, 9, 15
Central Florida GreeneWay . . .132-133, 135, 151, 153, 205
Central Florida Research Park 88-89, 92, 214
Chapin, Drew 191
Chapin, Linda 118, 125, 133-134, 212
Chapman and Son, Inc. 107
Chewning, Robert 43
Chickasaw Trail 30, 78-79, 82, 88, 91, 209
Chira, Lee 52, 55
Clayton, Charles 80
Clayton, Malcolm 80
Cocoa Beach 45, 47, 50
Codes of Ethics 128
Coin 157-158, 170

Cole, George 125
Coleman, Mel 36
Coleman, Vicki 4, 70
Colonial Drive 18-19
Commercial Bank of Winter Park 27
Common Cause 123
Cone Brothers of Tampa 34
Confiroute 172
Conway Road 15, 67
Cooper, James T. 47
Cooper, Jr., J. Fenimore 12, 14, 17, 28, 35, 69,
70, 79, 90, 98, 128
Coverdale and Colpitts 10, 11
Crawford, Bob 185
Curry Ford Road 15, 95-98, 100-101, 166
Cycmanick, Michael 74

D
Daetwyler 11, 13-14, 48, 50-51
Dailer, Gregory 4, 67, 123, 156, 171, 174, 178, 196, 212
Daniel, George 15-16
Dart Road 138, 150
Dayco Astaldi Construction 144, 146
Dean Road 73-77, 79, 83, 88, 91-93
DeLeuw Cather Engineering 91, 116-117
Denton, Keith 69
Department of Environmental Regulation (DER) . .73, 76,
77, 97, 104, 140-141
Department of Health and Rehabilitative Services 122, 124
Dial, William H. (Billy) 6
Dirt Trucks 93
Disadvantaged Business Enterprise (DBE) . . .127, 145, 164
Disney Development Co. 126
Dittmeier, John 71, 113
Diversacon, Inc. 32-38, 42, 59
Dore Wrecking Company 32
Dorman, Tom 69, 106, 136
Downtown Development Board 115
Doyle, Robert 45

Index of Key Words

Drage, Tom 115, 140, 184
Dunham, James 35
Dunnick, Chuck 190
Dura Stress, Inc. 144
Durek, Joe 129
Dyer, Riddle, Mills and Precourt 79, 83, 98, 126, 139-140, 195

E
E-PASS 150, 152, 156, 158, 160, 165-167, 169-170, 180, 198, 205, 209-210, 213
East Central Florida Regional Planning Council (ECFRPC) 7, 9, 19, 45, 138
East Orange Community Center 87
East-West Expressway 4, 18-20, 22, 23, 25-28, 31, 34-35, 38-43, 46, 50, 52, 54, 57, 60-64, 66-68, 73, 78, 80-83, 85, 87-88, 92, 94, 96, 98, 101, 103-104, 106-109, 130, 132-133, 159, 165, 167, 182, 205-207, 212-213, 216-217
Eastern Beltline 57-58
Eastern Beltway 62, 69, 78, 86, 95, 101-102, 109, 132-133, 135-136, 159
Eastern Bypass 61, 63, 64
Eby, Martin K. 144-145
Econlockhatchee River 14, 77, 81-84, 92, 94, 99, 141
Economic Development Commission 115
Edgewood 113-120, 185
Electronic Toll Collection and Traffic Management (ETTM) . . .157-159, 161, 163-164, 167-173, 175-177, 213
Embankment 93
Enforcement 157, 167-168, 176
Environmental 38, 49, 63, 76-77, 94-95, 97, 104, 133, 140-141, 154, 182, 189-191
Epcot Center 138
Equal Opportunity 122, 164
Ernest and Ernest 29
Erwin, Donald 4, 157, 171, 178
Evans, Donald 9, 10, 14

F
Fagan, Ron 163
Federal Aviation Authority (FAA) 51, 146
Federal Highway Administration 117-118, 206
Fiber Optic 169, 210-211
Figg & Muller 116
Figg Engineers, Inc. 117
Figueredo, Jorge 4, 130, 133-134, 148, 161, 163, 165-166, 169, 174, 192, 209
First Boston Corporation 13
Fletcher, Richard L. 4, 9, 14-17, 20, 22-30, 39, 43, 46-47, 56, 59, 77-78, 121, 213
Florida Central Railroad Line 198
Florida Department of Law Enforcement(FDLE) . . . .
. 124-125
Florida Department of Transportation (FDOT) . . . .
. . .4, 26, 29, 31, 39-40, 42-43, 49-54, 57-63, 66, 68-72, 89, 106-107, 109, 11, 114, 116, 118-120, 125, 132-135, 137-139, 147, 151-153, 155, 163, 169, 171-175, 177-180, 182, 184-186, 188, 190, 192, 195-196, 200-201, 203-204, 206-207, 209-210, 212-213,
Florida Development Commission 7, 47
Florida Division of Bond Finance 106
Florida East Coast Railroad 15
Florida Gas 48
Florida Power Company 95
Florida Ranch Lands 135
Florida Road 51, 71
Florida Statute 117, 140
Florida Technological University 7, 19
Florida Turnpike 6, 46, 59, 63, 65, 101, 103, 106-110, 139, 147, 150-152, 181, 187, 189, 195-196,
Florida Turnpike Authority (FTA) 10, 17-18, 44, 46-49
Forsyth Road 35-38, 85
Frederick, Bill 56, 115, 118
Freeman, Bob 187, 190
Freeman Truck & Equipment Company 107
Fullers Crossroad 198

Index of Key Words

G

GAI Consultants-Southeast	100, 163
Gage, Don	15
Gahr, Lloyd	9
Gaines, Harvey	29-32, 35, 43, 53-54, 56, 60, 79, 90, 98
Garland Avenue	29
Gary, Wilbur S.	69-70, 106, 125, 129, 133
Gee and Jenson	79, 87-89, 91
George Terry's Magnolia Ranch	15
Gibson, William L.	40
Gilliard, Charles	163, 173, 177
Glancy, John	160-161
Glatting, Jackson, Kercher, Anglin, Lopez & Rinehart	191
Gluckman, Casey	76
Gluckman, David	76
Goetz, Ludwig	76, 80
Goldenrod Road	14, 21, 73, 79, 81-82, 93, 165, 170, 208
Goldenrod Road Extension	207-209
Goldman, Eldon C.	14, 16-17, 25-27, 29, 34, 39, 59
Goldman Sachs & Co.	26
Good Homes Road	106, 108
Goodman, Martin	74, 77
Gordon, Vera	9
Gore Streets	116, 119
Graham, Bob	53-56, 71
Granite Construction Company	199
Grasshopper, Inc.	37
Gray, Jr., John L.	53-54, 70-71, 87
Greater Construction	76, 80
Greater Orlando Aviation Authority (GOAA)	50-51, 96, 136, 139, 144, 147-148, 207-208
Greater Orlando Chamber of Commerce	115
Greenbaum, Dan	130
Greene, James B.	27-29, 31, 34, 39-41, 43-44, 53, 56-60, 64, 69-70, 72, 77-78, 121, 132, 135-136, 213
Greiner Engineering	79, 117, 139-140, 191-192, 195
Ground-Breaking	15-16, 35, 82, 154, 186, 194, 198
Grovdahl, Dave	138

Gurney, J. Thomas	49, 59
Gustafson, Tom	184
Gwynn, David W. (Bill)	70-71, 77, 81, 83-84, 97, 114-115, 121, 123, 125, 128, 130, 155-156, 171

H

Hardage, Gerald	33
Hardaway Construction Company	80, 83, 100
Harrell, Robert S. (Bob)	69-70, 119, 125, 127, 133, 155
Haven, Robert	125
Hawkins, Robert C.	157, 171, 174
Henderson, Kaye	85, 183-184
Herndon Airport	29
Hewitt Construction Company	100
Hewitt Contractors	92
Hewitt, O.P.	9, 14
Hiawassee Road	22, 104-105, 110
High Point of Orlando	87-89
Highway Management Services	163
Hills	10, 15, 20-21, 23, 81, 108, 140
Hoffman, Elmo	35
Hoffman, Hendry, and Parker	35
Hoffner Road	208
Holland East Toll Plaza	67, 68, 158, 166, 167
Holland West Toll Plaza	41, 67-68, 167
Home Builders Association	31
Horizon West	187, 190, 200-201, 203
Houdaille-Duval-Wright	37
House Transportation Committee	53, 120
Houston, Nancy	133-134, 149, 186, 212-213
Howard, Needles, Tammen, and Bergendorff (HNTB)	10, 20-22, 24, 29, 39-40, 58-60, 140, 213
Hubbard Construction	15-16, 34, 50, 80, 83, 92-93, 144-147, 152-153, 197-198
Huckleberry	87-89
Hull, Alex	103-105
Hunter, Dan	28
Hunter, Pattilo, Powell and Carroll	28
Hunters Creek	136, 143

Index of Key Words

Hurricane Andrew 142

I
I-4 .... 10-11, 16, 18-19, 31, 34-35, 38, 45-46, 48-50, 57, 59, 65, 71, 111-112, 114, 117-119, 124, 132, 135-138, 150-151, 153-154, 169, 181, 183, 190, 199, 201-207, 210, 217
Intelligent Transportation Systems 210
Interchange Justification Report 117-118
International Drive 138-139
Interpark 172

J
J. E. Hill 37
Jackson, Timothy 89, 98
Jackson, Wiley N. 34, 49, 55
Jacksonville 7, 15, 20
Jennings, Toni 184, 201
Jernigan, Alex 157
John Young Parkway 34, 43, 139, 142, 144, 206
Johnson, Ray 13
Joint Partnership Agreement (JPA) 203
Jones, Fred 53-54

K
K &.L Contractors 82
Kaiser Engineering 91, 107
Kantor, Hal 96
Kelly, John 29, 39, 59
Kelly, Tom 105
Kerce, Joe 178
Ketelson, Kenneth 30
Kiewit 172
Kimley-Horn and Associates, Inc. 124-126, 195
Kirby Smith Groves, Inc. 143
Kirk, Claude R. 25, 46, 48-49,
Kirkman Road 104-106, 109, 205-206
Kissimmee 150

Kunde-Sprecher & Associates 195
Kunde, Sprecher and Yaskin 108

L
Lacy, Brent 113
Lake Barton Road 15, 21,
Lake County 103, 141, 211-212, 216
Lake Holden 113-114
Lake Jessup 135, 151-153
Lake Lucerne 6, 29, 31, 34, 41, 214
Lake Mary Road 151
Lake Nona 136, 207
Lake Sherwood 104, 106
Lake Underhill ... 21, 23, 28, 30, 32, 38, 87, 95, 98-99, 206
Lakeland 34, 120
Lamar, Lawson 24
Land, John 86, 90, 94
Landstar Road 139, 143-144
Langford, Carl 6, 35
Lawson, James Forest (Sonny) 143
Lawther, Wendell 173
Lease Purchase Agreement 66
Lee Vista 96
Lee Vista Boulevard 96, 99-100, 192, 207-209
Leecon and Craggs & Phelan 37
Lewis, H. E. 15
Lewis, Thomas 126
Livingston Meadows 104
Lochner, H. W. 140, 195
Lockheed-Martin 5
Long, Inez J. 133-134, 149, 164, 200
Long Range Expressway Plan 62-63, 73, 87, 95, 103, 109, 111, 136, 182, 215-216
Lowndes, Drosdick, Doster, Kantor and Reed 126
Lynx 207

M
M. G. Lewis and Company 66
Mainline Plaza 179, 198

Index of Key Words

Maitland Boulevard Extension 181, 186, 190
Maitland Interchange 57
Mandell, Robert 76, 125, 132-134, 149,
                                                          161-162, 164, 212
Martin Company 82-83, 109, 126
Martin Paving Company 82-83, 109, 126
Martin, Robert D. 126
Martin-Marietta 5, 48, 87, 172
Martinez, Bob 82, 85-86, 94, 122-125, 183-85
Martinez, Mel 196, 200, 212
Marvin, Eddie 151
Marwick, Peat 123
Mazzillo, Darleen 4, 70, 86
Melbourne 47
Metric Constructors, Inc. 108
McCarey, Pat 119
McCormick Road 198-199
McCoy Jetport 6-8, 10-11, 47, 49-50
McCoy Road 11, 13-14, 48, 50-51, 55
McKelvy, William B. 4, 71-72, 80-81, 101,
                                                  141, 145, 148, 163, 178
McNamara, Joseph 71
McNulty, Clifford 47
Meadow Woods 136
Mercado Drive 31, 34, 37
Metropolitan Life Insurance 35
Metropolitan Planning Organization (MPO) . . . .
. . . . 61, 73, 111, 181
Metropolitan Transportation Authority 74
Mica, John L. 154
Michigan Street 111, 113, 117, 119
Miller, Miller, Sellen, Einhouse, Inc. 136
Mills Avenue 42-43, 68
Mills, Russell 126
Minority Business Enterprise Office (MBE) 126-127
Misener Marine 38
Missing Link 154
Mizo, Hill and Glass 92
Mizo-Hill, Inc. 82

Moore, Cecil 119
Moreland Altobelli Associates 163
Morrison Knudson 172, 177
Moss Park Road 145
Mueller, Edward 26-27, 34
Muszynski, James 112, 122-123

N

Narcoosee Road 11, 18, 139, 145-146
National Aeronautics and Space Agency (NASA) . . . 19, 45
Naval Training Center 19, 21
Needles 10, 140, 213
Norrell Temporary Services 173
Northeastern Beltway 73-76, 79-82, 84-85, 73,
                                   75, 80-81, 85-86, 90, 92, 97-98
Northwestern Beltway 181

O

O. R. Colan and Associates 31
Oak Ridge High School 116
O'Brien Construction Co. 37
Ocoee 129, 182, 186-187, 189-190, 196, 216
Office of Toll Operations (OTO) 155, 178, 209-210
Ogden, Robert 11, 14
Old Winter Garden Road 42, 108
Orange Avenue 5-6, 11, 48, 111, 113-114,
                                                        120, 137, 139, 140
Orange Blossom Trail 11, 41, 48, 68, 111, 136, 143
Orange Concrete 37
Orange County Commission 8, 10, 14, 28, 36, 47-48,
                               51, 53-54, 56-58, 62, 69-70, 75, 87,
                               103, 117, 129, 136, 141, 190, 194
Orange County Democratic Executive Committee 115
Orange County Homeowners Association 115
Orange County School Board 61
Orange Lake 201, 203
Orange Ridge 76
Orlando Area Chamber of Commerce 9, 13
Orlando Business Journal 112, 159

Index of Key Words

Orlando Central Business District, Inc. 29
Orlando Federal Savings and Loan Association 31
Orlando Housing Authority 31
Orlando International Airport 4, 48, 50, 53, 55,
    96, 102, 111, 137, 142, 145, 147-148, 207, 214, 217
Orlando Neighborhood Improvement Program 31
Orlando, Seminole, Osceola Transportation Authority
    (OSOTA) 57-58
Orlando Sentinel 6, 28, 45, 59, 76, 91, 103-104,
    112-114, 122, 136, 165-166, 184, 186
Orlando Urban Area Transportation Study (OUATS) ....
    ....57, 61, 181, 190
Orlando Utilities Company 144
Orlando-Orange County Expressway Authority (OOCEA)
    ....4, 5, 7-18, 20, 22, 24, 26-37, 39-48,
    51-63, 66, 81, 83-90, 93-100, 103-109, 111-142,
    146-149, 154-165, 167-197, 199-201, 204-213, 215-218
Orlando-Winter Park Board of Realtors 31
Orlo Vista 104
Orsino Causeway 45-47, 49-50
Osceola County 40, 57-59, 132, 136-139, 150,
    169, 181, 185, 188-190, 200, 202-203, 211
Osceola Parkway 150, 169
Outer beltline 35, 39-40, 43, 57, 59
Oviedo 151, 169

P
Parramore 31, 34, 41
Parsons Brinkerhoff 81, 172
Parsons, Brinkerhoff Construction Services 81, 172
Part A 120, 138-140, 181-182, 184-186,
    189, 191-202, 205, 211-212, 215-216
Part B 138-140, 181, 186, 189-90, 204
Part C ...11, 181, 186-189, 191-193, 195, 199-205, 215-216
Patrick AFB 45
Patten, Dale 29, 39, 59
Paulsen, Bob 132
Pebbles, Willard 14, 16-17, 20, 24-25, 41, 45
Pepin, John 88, 90

Perry & Lamb 133
Peterson, W.C. 15
Pfeiffer, Frederick 118
Pickett, Paul 39
Planning and Zoning Commission 76
Poe, Ralph 27-28
Polk County 120, 181, 185
Poorbaugh, William 39-40, 59
Pope, David 163, 174, 177-180
Port Canaveral 45-46
Post, Buckley, Schuh & Jernigan, Inc. (PBS&J) ....
    ....4, 51-54, 62, 64, 70, 73-75, 79, 89, 95-97,
    111-112, 125-126, 128-129, 132, 156-157,
    171, 174, 178-179, 185-186, 205, 212
Powers Drive 21
Prime Design, Inc. 91, 98
Professional Engineering Consultants/ W.K. Daugherty,
    Inc. (PEC) 103, 106, 116
Public Financial Management 67
Pugh, Irby 76
Pugh, James 200, 212-213
Pustelnyk, Steve 4, 163, 167

Q
Quesinberry, Jack 190

R
Ramp 35, 42, 54, 67-68, 82, 99-100, 129-130,
    145, 149, 152, 157, 160, 166, 179, 198, 206-208, 216
RCH & Associates 156-157, 171, 174
Red Bug Lake Road 152-153, 169
Reece, Phil 4, 56, 69, 78-79, 85-86, 103, 106, 136
Reedy Creek Improvement District 150, 159, 201-203
Refunding 66, 131
Regional Expressway Authority 74, 83
Reinforced Earth, Inc. 101
Request for Proposals (RFP) 158, 171
Revell, Walter 39, 59
Revenue Bonds 10, 150-151

Index of Key Words

Rex, Jr., Charles W. 47
Reynolds, Smith and Hills 15, 20, 108
Rich, A. Wayne 4, 132, 134, 149, 188, 190, 200, 236
Right-of-way 13, 19, 21, 24-26, 29-34, 37, 43, 48-51,
    57, 60, 63-64, 71-72, 76, 79-80, 82-85, 87, 89-91,
    93, 95-101, 105, 107-108, 113-114, 116, 127, 135-136,
    138, 142-143, 145, 147, 151, 153-154, 187, 189, 190,
    192-196, 199-201, 203, 207-209, 211, 220, 226-227
Riverwood Village 95
Rizza, Myrtle 43, 70
Robaina, Tony 96
Rogers Group Contractors 92
Rosalind Avenue 67, 206
Rose, William N. 53
Ross, Tom 133, 162, 178, 193, 236
Rouse Road 75, 91, 93
Rushing, John 24, 29, 35, 43
Russo, Tom 89

S
S. E. Montgomery Trucking 37
S. R. 15 146
S. R. 50 ....6, 12, 14, 18-22, 31, 34, 36-37, 42, 63,
    67, 73, 7879, 81-83, 85, 89, 91-92, 104-106,
    108-109, 129,181, 194, 196-198, 212, 216, 217
S. R. 426 .....63, 73, 79, 81, 83, 86, 109, 135, 138, 151-153
S. R. 434 152-153
S. R. 436 .....15, 21, 36, 48, 50, 62, 68, 111, 147, 182, 217
S. R. 520 ....11, 13, 14, 50, 57, 217
S. R. 535 ....132, 135, 138-140, 144, 150-151, 169, 204
SAIC 159-165, 172, 218
Sand Lake Road 11, 16, 18
Sanford 135, 153-154, 183, 190
Scharlin, Howard 96, 99
Schweitzer, Nils 41
Scientific Applications International Corporation ....
    ....156, 159, 219
SEACOR 155
Seidel Road 203-204

Sellen, Jim 136
Seminole County 9, 59, 60, 62, 73, 74, 79, 83-84,
    86, 109, 132, 135, 153, 169, 181, 185, 217
Seminole County Expressway 74, 151, 135, 184
Seminole County Expressway Authority (SCEA) ....
    ....4, 74, 79, 82-84, 151, 152, 154
Seminole One 86, 138, 152
Seminole Springs 141
Sentinel Star 40, 44
Sierra Club 30
Silva, Nathan 191
Silvia Lane 29
Simon, Susan 4, 70, 71, 86
Sindler, Bob 194
Singleton, Sidney 11, 14
Skau, Ed 115
Slack, Teresa 212
Slaughter, Brantley 104-105, 107
Sloan Construction Co. 37
Smith, Barney and Company 27
Smith, Jim 148-149
Smith, Sandy 119
Snyder, Mike 200, 212, 214
South Florida Water Management District 140
South Orange Community Council 112, 122-123
South Orlando Sun 129
Southchase 136
Southchase, Ltd. 143
Southeastern Beltway 95-98, 100-101, 106-107, 145
Southern Connector 109, 116, 120, 128, 131, 133,
    135-150, 159-162, 164-166, 173-174, 180, 184, 193, 197
Southern Connector Extension 132, 135, 139, 150, 153
Southland Construction Company 92, 100
Southwest Beltway 181, 189, 199
Spessard Lindsay Holland 34
Staley, Tom 194
State Board of Administration 12
State Road Department (SRD) 4, 7, 9-10, 12-13,
    15, 18, 20, 23-25, 46-49, 57

Index of Key Words

Steinmetz, Phillip D. 31
Stevens, R.W. 50
St. John's River 48, 50, 97
St. John's River Water Management District 73, 140
Story Road 198
Structures, Inc. 34, 37
Stuart, Jr., George 53
Summerlin Avenue 34
SunPass 209, 210
Sunshine Amendment 122, 125, 128
Sunshine State Parkway 46
Sverdrup & Parcel Associates, Inc. 91
Swann, Richard 39-40, 59
Sylvester, Charles C. 4, 70-72, 79, 89, 98, 98, 143, 212, 213
Symmes, Jewel 71
Syntonic, Inc. 156, 159, 163

T
T. G. Lee 96
Tallahassee 76, 124, 185
Talton, John H. 14, 27-28, 39-40
Tammen 10, 21, 140, 213
Tangerine Bowl-Tinker Field 29
Telos Corporation 174
Thompson, Emerson 112
Titusville 47, 50
Toll Gates 68
Toll Increases 64, 67, 129
Toll Operations Contractor (TOC) 175-178
Toll Plaza 11, 14, 16, 41-42, 50-52, 55, 67-68, 80, 82, 84, 92, 93, 100, 108, 139, 140, 142, 144-145, 152, 158, 165-166, 175, 179, 180, 198, 198-199, 208, 212
Toll Roads 46, 60, 106, 109, 121, 129, 130, 155, 181, 188, 213
Tradeport Drive 51
Treadway, Lou 56, 70
Tri-County League of Cities 115
Trinity Industries, Inc. 144

Trux, Bob 15
Tuskawilla Road 115

U
U. S. 441 ....6, 11, 13-14, 63, 139, 143-144, 182, 186, 193, 195, 199, 206, 211, 216
U. S. Navy 51
URS Consultants 177
Union Park 21, 35
United Infrastructure Corporation 177
University Boulevard 75-77, 79, 83
University Boulevard Coalition 74-77
University of Central Florida 75, 88

V
VSL Corporation 144
Valencia College Lane 79, 82
Valencia Community College 74
Vandergrift, Scott 190, 194
Varela, Patricia 4, 70, 167
Vehicle Enforcement System (VES) 176
Violation 68, 106, 157, 167-168, 176-177
Vollmer and Associates 67, 85, 89, 111, 129, 130, 186
Volusia County 10, 217
Voorhies, Trindle-Nelson (VTN) 24, 57

W
WBE 127
Walt Disney 17, 19, 45
Walt Disney World 45, 187, 203, 213
Walsh, Kevin 126
Wasleski, Victor F. 14
Watermill 76
Watson and Company 24
Watts, Ben 70, 125
Webber, Paine 69, 186, 196
Webster, Daniel 184, 196, 201
Wekiva 154, 181-182, 190

Index of Key Words

West, Betty Jean (B.J.) 69-70, 93-94, 116, 121-125, 130
West Orange Trail 198
Western Extension 63-65, 98, 100-101, 103-110,
132, 182, 216
Western Beltline 57
Western Beltway 63, 103, 109, 119120, 154,
181-196, 200, 202-205
Western Bypass 61, 63-65
Wetlands 73, 77, 97, 140-141
White Construction Company 152
White, Lynn 178
Wilbur Smith and Associates 20, 58-60, 74, 100, 140
Wilson Construction Co. 37
Windshield-Mounted 210
Winter Garden 129, 181182, 186-187, 189-190,
192, 194, 199
Winter Park 14, 27, 28, 31-32, 34, 43, 114
Winter Springs 151
Worrall, Harold 4, 130, 133-134, 149, 156,
160-164, 178, 193, 201, 207
Wuestefeld, Norman 21, 60

X

Y

Z
Zipperly Hardage and Associates (ZHA) 78, 81, 86,
91, 98, 107, 116, 128, 144, 163